# Drought and
# Natural Resources Management
# in the United States

W0113663

# Drought and Natural Resources Management in the United States

## Impacts and Implications of the 1987–89 Drought

William E. Riebsame,
Stanley A. Changnon, Jr.,
and Thomas R. Karl

Routledge
Taylor & Francis Group

LONDON AND NEW YORK

First published 1991 by Westview Press, Inc.

Published 2018 by Routledge
52 Vanderbilt Avenue, New York, NY 10017
2 Park Square, Milton Park, Abingdon, Oxon OX14 4RN

*Routledge is an imprint of the Taylor & Francis Group, an informa business*

Library of Congress Cataloging-in-Publication Data
Riebsame, William E.
    Drought and natural resources management in the United States : impacts and implications of the 1987–89 drought / by William E. Riebsame, Stanley A. Changnon, Jr., and Thomas R. Karl.
        p.   cm. — (WVSS in natural resources and energy management)
    Includes bibliographical references and index.
    ISBN 0-8133-8026-X
    1. Droughts—United States—Management.   2. United States—Climate.   I. Changnon, Stanley Alcide.   II. Karl, Thomas.   III. Title.   IV. Series.
QC929.D8R52   1991
363.3'492—dc20                                                              90-40603
                                                                              CIP


ISBN 13: 978-0-367-01547-3 (hbk)

ISBN 13: 978-0-367-16534-5 (pbk)

# Contents

# Figures and Tables

# Acknowledgments

An endeavor such as this is not possible without many people's indispensable assistance. Valuable advice and reviews of draft material were provided by Peter Lamb—Illinois State Water Survey, Mike Helpa—U.S. Army Corps of Engineers, Alan Hecht—Environmental Protection Agency, and Don Despain and Phil Perkins—Yellowstone National Park. Their help is greatly appreciated. Special thanks also go to Joyce Changnon for review and encouragement.

We take responsibility for the interpretation, conclusions, and recommendations, and our conclusions do not necessarily reflect those of the many agencies, such as the National Park Service, U.S. Army Corps of Engineers, U.S. Department of Agriculture, and National Oceanic and Atmospheric Administration, which played important roles in responding to the 1987-89 drought.

Special thanks go to Richard Heim of the National Climatic Data Center (NCDC), who is indeed a valuable national resource for climate information. Additional information was provided by Cameron Johnson—U.S. Forest Service Fire Sciences Laboratory, Dick Reinhardt—Western Regional Climate Center, Ken Hubbard—High Plains Climate Center, Ken Kunkel—Midwestern Climate Center, John Purvis—Southeastern Regional Climate Center, Warren Knapp—Northeastern Regional Climate Center, Moulton Avery—Center for Environmental Physiology, Dave Morton—Natural Hazards Center, and James Laver—Climate Analysis Center.

The research for the Mississippi and Atlanta case studies was conducted as part of the applied studies program of the Midwestern Climate Center, which is supported by NOAA/NCPO (Grant COMM NA8711-D-CP119) and state of Illinois funds.

The Yellowstone and North Dakota case studies were supported by the Natural Hazards Center, as was the overall editing and publishing process. David Diggs conducted the survey of North Dakota farmers reported in Chapter 5 as part of a "Quick Response" grant from the Natural Hazards Research and Applications Information Center (NHRAIC). The center is supported by nine federal agencies through the National Science Foundation's (NSF) Earthquake Mitigation and Natural and Man-Made Hazards Mitigation programs.[1] Bill Anderson and Eleanora Sabadell at NSF have been especially thoughtful and conscientious supporters of hazards research.

Scott Miller and Ronald Baldwin of the National Climatic Data Center and Dennis Ehmson of the University of Colorado Office of Contracts and Grants prepared most of the figures. Additional editorial assistance was provided by Illana Gallon and Wendy Hessler of the Institute of Behavioral Science, University of Colorado.

Finally, we owe many thanks to Sylvia Dane, Natural Hazards Center editor, for her hard work, good professional sense, and patience with this book. It would not have been published without her help.

*William E. Riebsame*
*Stanley A. Changnon, Jr.*
*Thomas R. Karl*

### Notes

1. The center is funded by the National Science Foundation, the Federal Emergency Management Agency, the National Oceanic and Atmospheric Administration, the U.S. Geological Survey, the Tennessee Valley Authority, the U.S. Army Corps of Engineers, the Environmental Protection Agency, the National Institute of Mental Health, and the U.S. Department of Agriculture, Soil Conservation Service.

# 1

# Introduction

The 1987-89 drought was a signal event in the evolving inter-relationships among climate, natural resources management, technology, and society in the United States. Over half of the country experienced severe to extreme drought by midsummer of 1988 (Figure 1.1). Losses upward to $39 billion illustrate the continuing, perhaps growing, vulnerability of many natural resources and economic sectors to drought and other climate fluctuations.

Despite decades of crop breeding, water system development, and other improvements in climate-sensitive technologies, the drought demonstrated that the simple lack of "normal" rainfall still provokes serious disruptions in agriculture, water supply, transportation, environmental quality, and other areas. It can affect the health and well-being of millions of people and evoke billions of dollars in government aid.

Several disturbing aspects of the drought presage continuing problems in the future. First, many natural resources managers were slow to recognize the significance of the event, and information on the drought's development and impacts was not disseminated and applied to critical decisions in a timely fashion. Natural resources managers often did not know how sensitive their systems were to drought, what impacts to expect from a severe drought, or the full range of options available for adjusting their activities. Despite a decade of growing interest in the social and economic impacts of climate fluctuations, codified in the 1979 National Climate Program Act, the nation remains ill-prepared to cope with unusual climate conditions.

The drought did evoke some successful responses, and lessons from past droughts were profitably applied in some cases. Indeed, those

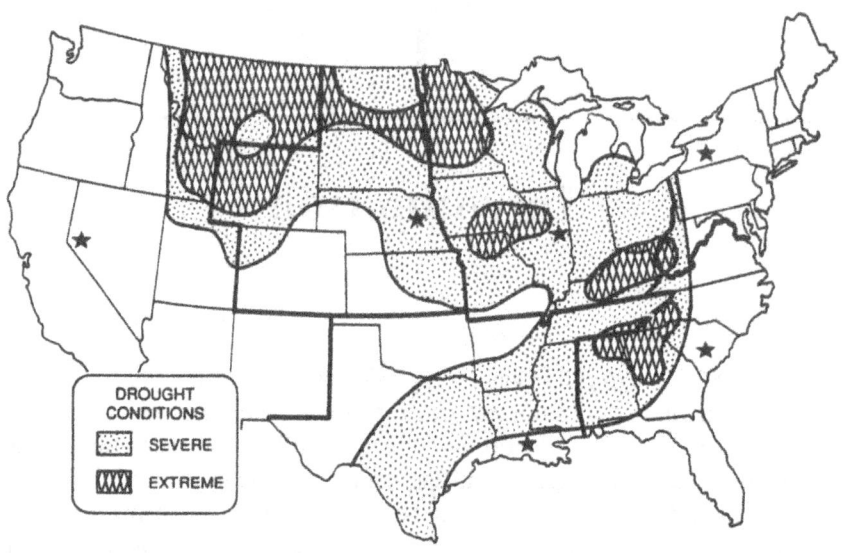

**Figure 1.1**
The drought's severity on June 15, 1988, as indicated by the Palmer
Drought Severity Index (*Source*: National Climatic Data Center)

successes provide the rationale for this book: they indicate a greater
potential for reducing drought impacts than was observed, especially
during the height of the drought in 1988. By diagnosing this case, and
placing it in the context of the evolving relationship between climate
and society, we seek to point the way toward improved drought
management in the future, as well as to better illuminate the path to
reduced overall climate vulnerability.

## The Drought's Legacy

The 1987-89 drought in the U.S. had several subtle effects on the
way both the public and private sectors operate under conditions of
climate stress and uncertainty. It raised the nation's consciousness about
the role climate plays in the management of natural resources and
illustrated how social vulnerability to natural hazards can change
through time due to economic and policy changes. Finally, the link
between the 1987-89 drought and longer-term climate change caused

by the Greenhouse Effect, made by a few scientists and widely discussed by the news media, turned the drought into a symbol of potential climate disasters in the future.

The need then, and what we attempt with this book, is to identify the successes and failures of 1987-89, first broadly and then through detailed case studies focused especially on the summer of 1988, and next to examine what this diagnosis implies for the management of future droughts and other climate fluctuations. We also identify a set of abiding, underlying problems in the management of climate-sensitive natural resources: problems of bias and poor information processing will likely emerge again during the next drought. These problems presage more fundamental problems in dealing with other climate hazards, such as global warming.

## The Drought Hazard in the United States

The 1987-89 drought was not quite the worst drought in modern times (the period 1934-36 was drier), nor was it the classic "natural disaster" to which most of the nation's emergency planning and preparedness efforts are geared. Rather, it was a mixture of cumulative and catastrophic events that interacted with social vulnerabilities to cause loss and hardship and was partially offset, in some cases, by economic gains.

Droughts are a normal part of the earth's climate, and essentially no geographical location is immune to occasional spells of unusually dry and hot weather. They have occurred throughout human history in most parts of the world, occasionally precipitating famines such as those witnessed in Sahelian Africa, and droughts will continue to occur in the future. The specific problems caused by droughts stem from their impacts on a wide array of natural systems (Figure 1.2) and climate-sensitive human activities. The hazards posed by droughts are best defined in terms of human management and use of natural resources—water, forests, and agricultural lands—whose productivity is affected by fluctuations in precipitation and temperature. These physical interactions produce impacts that ripple through the social structure as physical, economic, social, and even political effects (Figure 1.3).

## The Country's Evolving Drought Hazard

Droughts have repeatedly reduced social well-being and made headlines in U.S. history (Warrick, et al. 1975). Drought in the early 1890s dramatically curtailed national agricultural expansion, especially in the midcontinent, and evoked fears of nationwide famine. Geographers Warrick and Bowden described the 1890s drought as:

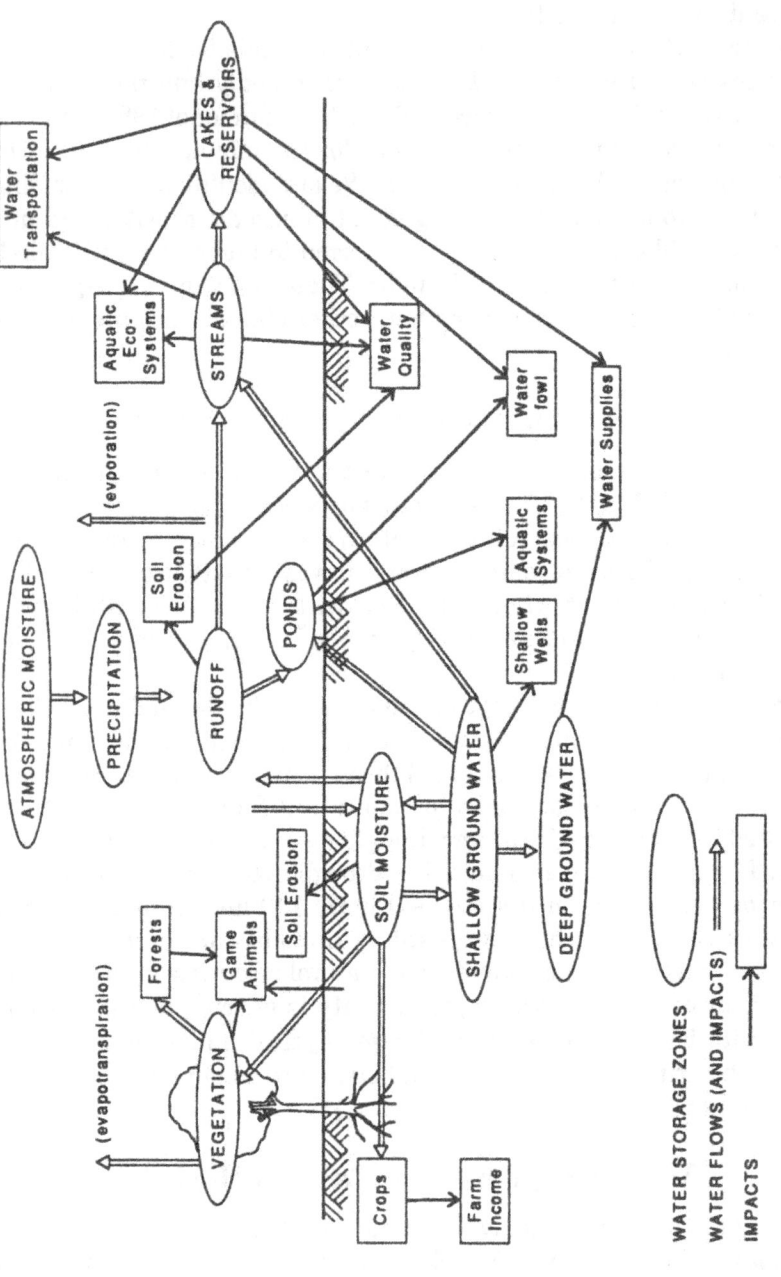

**Figure 1.2**
Environmental elements affected by droughts and their related impacts (*Source:* Changnon and Easterling 1989)

a disaster. After a number of years in which opportunistic agricultural turalists flooded into the largely uncultivated margins of the frontier, the region experienced a net displacement of nearly three hundred thousand people. Many areas lost between half and three-quarters of their population (1981: 126).

Harlan Barrows referred to it as the "first great crushing defeat" of the American farmer (1962: 231).

Vulnerability to drought took on new dimensions after the turn of the century, as greater investment was made in public works for agriculture, water supply, and transportation. The U.S. Department of Agriculture (USDA) pushed for irrigation, new crop breeds, and "dry farming" of the semiarid regions, a pseudo-scientific method for preserving soil moisture (Hargreaves 1956). Public funds were invested in water resources for transportation, domestic supply, and the western irrigation systems that were to provide a capstone to the country's agricultural development.

However, infrastructural investments and new technologies did little to lessen the impacts of the next major drought in the mid-1930s, the so-called "Dust Bowl" years. More than any drought in U.S. history, the 1930s imprinted themselves on the American collective consciousness

Figure 1.3
Range of possible social consequences of drought (*Modified from* Warrick, et al. 1975)

(Riebsame 1986): 1934 and 1936 were the two driest years in the recorded history of U.S. climate, dryness that precipitated one of the country's worst natural disasters (see Chapter 2). The economic system was especially sensitive in the 1930s—following the stock market crash of 1929—and the combination of social vulnerability and climatic extreme produced hardships unmatched since that time. Nowhere was this more evident than in the American Great Plains, where thousands of tenant farmers were "droughted out" (Hurt 1986), and soil from the agricultural lands of the midcontinent blew in huge dust storms, occasionally settling on the eastern seaboard where policy makers were fashioning a new era of natural resources and hazards management (Great Plains Committee 1937; Worster 1979). The 1930s also stand out in the drought response plans of most natural resources systems. The run of dry years from 1931 through 1936 is typically used as the "benchmark" drought, or the "drought of record," for water management systems throughout the country (e.g., Russell, Arey, and Kates 1970).

Much was accomplished following the 1930s to lower the country's future vulnerability, both urban and rural, to drought: reservoirs were built or enlarged and domestic water systems improved with public funds, farm policies were changed drastically, new insurance and aid programs emerged, and some of the most sensitive agricultural lands were taken out of production (Hurt 1986).

When major drought emerged again in the mid-1950s, affecting the Southwest and midcontinent, impacts were less severe: fewer farms failed, people did not go hungry, and urban centers continued to receive water supplies—even where the drought was as severe as that in the 1930s (Warrick, et al. 1975; Bowden, et al. 1981). But the next major U.S. drought affected a region unused to, and unprepared for, dry spells: in the 1960s it was the Northeast's turn, and many urban water systems that managers felt were reliable either failed outright or were on the verge of failure when the rains returned. Most New England and mid-Atlantic water systems functioned only through extraordinary measures and unprecedented water conservation (Russell, Arey, and Kates 1970). This ushered in a new era of drought response: rather than trying to assure ever more water during droughts, managers focused on more flexible water systems operation and reduced water demand—a strategy elaborated into a new paradigm of water management in California during the western U.S. drought of 1976-77.

It was in 1976-77 that the contemporary pattern of governmental response to drought emerged (Rosenberg 1980; Wilhite 1983). The 1976-77 drought affected large portions of the far West, Great Plains, and upper Midwest, causing direct losses of $10-$15 billion. As the drought entered its second year, President Carter declared a drought emergency and, in February 1977, appointed a national drought coordinator. An Interagency Drought Coordination Committee was formed in April that eventually designated 2,145 U.S. counties (two-thirds of the country) as drought disaster areas. Some 40 separate drought relief programs, costing an estimated $8 billion, were administered by 16 federal agencies (Wilhite, et al. 1986). But, lack of coordination and poor drought monitoring meant that relief efforts were inefficiently implemented. Indeed, Wilhite, a drought policy expert, concluded that:

> The United States respond[s] to drought through crisis management rather than by risk management. This was true not only in the mid-1970s but also in previous episodes of widespread and severe drought ... Reaction to crisis often results in the implementation of hastily prepared assessment and response procedures that may lead to ineffective, poorly coordinated, and untimely response (1987: 428-29).

On the heels of the mid-1970s drought, and following the recommendations of several drought research and policy workshops held in the late 1970s and 1980s (e.g., Rosenberg 1980; Wilhite, et al. 1987), a few states, such as Colorado, Nebraska, and Kentucky, developed comprehensive drought response plans. California, which was severely affected by the 1976-77 drought, made great progress in linking drought monitoring, assessment, and response (California Department of Water Resources 1978). Yet, as the events of 1988 illustrate, many state and federal governments continued to respond with the ad hoc crisis management strategies that were criticized after the 1976-77 drought.

### How Well Do We Manage the Drought Hazard?

Droughts are a difficult natural hazard to manage. They emerge slowly, creeping up on resource managers and developing from seemingly insignificant beginnings—a dry spell of a few days or weeks—into natural events that affect every element of the environment and society (Changnon 1987; Changnon and Easterling 1989). Hazard analysts have argued that drought should receive greater priority in the nation's plans for natural disasters, yet it remains neglected. A panel reviewing the National Climate Program in 1986 concluded that:

> The apparent lack of drought planning at the federal level suggests that the next major U.S. drought will again evoke an inefficient and poorly coordinated response, as did the droughts of the 1970s (National Academy of Science 1986: 14-15).

We found evidence in the case studies described in this book that the 1987-89 drought reflected some adaptive changes within agencies and institutions involved in natural resources management. But, we also found much that did not change, and uncovered little reason to expect that the effects of the next major U.S. drought will be better handled.

The severity of the drought, especially in summer 1988, was poorly recognized early on despite climatological indications during the fall and winter of 1987 and spring of 1988 that it was evolving into a severe and dangerous event. In other cases, poor coordination between people with key climate data and people needing to apply the data to important decisions led to delays and inefficiencies. When the various agencies and institutions began disseminating drought information in a coordinated fashion during June 1988, it was quite skillfully produced and well-received by decision makers sufficiently sophisticated to recognize its value.

The 1987-89 drought was also different from past events because it was linked to a broader and more enduring environmental problem—global warming caused by the so-called Greenhouse Effect. By virtue of its severity, timing, public attention, and link to broader environmental problems, the drought evoked fundamental concerns about sustainable natural resources management in the United States. The resulting institutional and policy impacts will reverberate throughout the social and political systems for some time to come.

## Goals and Structure of This Book

Our purpose here is to diagnose the 1987-89 drought,[1] both physically and socially, and to draw a prognosis and prescription for the future management of climate-sensitive natural resources—management that can reduce the impacts of droughts and other climate fluctuations. First, we examine the climatological nature of the drought (Chapter 2) and assess the broad range of impacts across social and economic sectors (Chapter 3).

Next, we offer four case studies that combine different aspects of drought effects and resources management: water and transportation management in the Mississippi River system (Chapter 4), dryland farming in the northern Great Plains (Chapter 5), metropolitan water supply provision in Atlanta (Chapter 6), and wildfire management in Yellowstone National Park (Chapter 7). For each case, we describe how the drought transformed natural resources management into crisis management, and explore in more detail the response and decision-making processes. We then assess what the drought indicates more generally about natural resources managers' abilities to cope with future unusual climatic conditions.

In the final chapter, we evaluate responses to the drought more broadly, identifying the surprises, problems, and solutions that emerged. We then lay out the components of improved drought management, including steps that should be taken before the next significant drought affects the nation. We conclude by speculating on what the impacts of drought indicate about broader climate and society interaction in the U.S. and the future consequences of climate fluctuations, both natural and human-induced.

## Notes

1. We will refer to the "1987-89 drought," recognizing that the dry spell, which had started as early as 1984 in the Southeast, took on national dimensions in 1987, became a truly severe nationwide drought in the summer of 1988, and continued into 1989. Most of our discussions of specific impacts and responses will focus on the summer of 1988.

## References

Barrows, H.H. *Lectures on the Historical Geography of the United States as Given in 1933.* Edited by W.A. Koelsch. Department of Geography Research Paper No. 77. University of Chicago. 1962.

Bowden, M.J., R.W. Kates, P.A. Kay, W.E. Riebsame, R.A. Warrick, D.L. Johnson, H.A. Gould, and D. Weiner. "The Effect of Climate Fluctuations on Human Populations: Two Hypotheses." In *Climate and History.* Edited by T.M.L. Wigley, M.J. Ingram, and G. Framer. 479-513. Cambridge, UK: Cambridge University Press. 1981.

California Department of Water Resources. *The 1976-77 California Drought: A Review.* Sacramento. 1978.

Changnon, S.A., Jr. *Detecting Drought Conditions in Illinois.* Circular 169. Champaign: Illinois State Water Survey. 1987.

Changnon, S.A., and W.E. Easterling. "Measuring Drought Impacts: The Illinois Case." *Water Resources Bulletin* 25: 27-42. 1989.

Great Plains Committee. *The Future of the Great Plains.* House of Representatives Document No. 144, 75th Congress. Washington: U.S. Government Printing Office. 1937.

Hargreaves, M.W.M. *Dry Farming in the Northern Great Plains: 1900-1925.* Cambridge, Massachusetts: Harvard University Press. 1957.

Hurt, R.D. *The Dust Bowl.* Chicago: Nelson-Hall. 1981.

——. "Federal Land Reclamation in the Dust Bowl." *Great Plains Quarterly* 6: 94-106. 1986.

National Academy of Sciences. *The National Climate Program.* Washington. 1986.

Riebsame, W.E. "The Dust Bowl: Historical Image, Psychological Anchor, and Ecological Taboo." *Great Plains Quarterly* 6: 127-136. 1986.

Rosenberg, N.J., ed. *Drought in the Great Plains: Research on Impacts and Strategies.* Littleton, Colorado: Water Resources Publications. 1980.

Russell, C.S., D.G. Arey, and R.W. Kates. *Drought and Water Supply.* Baltimore: Johns Hopkins Press. 1970.

Warrick, R.A., and M.J. Bowden. "The Changing Impacts of Drought in the Great Plains." In *The Great Plains: Perspectives and Prospects.* Edited by M.P. Lawson and M.E. Baker. 111-137. Lincoln: University of Nebraska Press. 1981.

Warrick, R.A., P.B. Trainer, E.J. Baker, and W. Brinkmann. *Drought Hazard in the United States: A Research Assessment.* Monograph No. 4. Boulder: Natural Hazards Research and Applications Information Center, University of Colorado. 1975.

Wilhite, D.A. "Government Response to Drought in the United States: With Particular Reference to the Great Plains." *Journal of Climate and Applied Meteorology* 22: 40-50. 1983.

——. "The Role of Government in Planning for Drought: Where Do We Go From Here?" In *Planning for Drought: Toward a Reduction of Societal Vulnerability.* Edited by D.A. Wilhite, W.E. Easterling, and D.A. Wood. 426-44. Boulder, Colorado: Westview Press. 1987.

Wilhite, D.A., N.J. Rosenberg, and M.H. Glantz. "Improving Federal Response to Drought." *Journal of Climate and Applied Meteorology* 25: 332-342. 1986.

Wilhite, D.A., W.E. Easterling, and D.A. Wood, eds. *Planning for Drought: Toward a Reduction of Societal Vulnerability.* Boulder, Colorado: Westview Press. 1987.

Worster, D. *Dust Bowl: The Southern Great Plains in the 1930s.* New York: Oxford University Press. 1979.

# 2

---

# A Climatological Perspective
# of the 1987-89 Drought

How does the 1987-89 drought compare to previous droughts in the U.S.? Was it unprecedented, and if so, in what respects? Are droughts becoming more frequent in the U.S.? This chapter addresses these and other questions that help us understand and place the drought in perspective. We do this by focusing on the temporal development of the drought from regional and national perspectives and by comparing its evolution and intensity with past droughts. We will see, however, that there are several ways to characterize droughts, and that some approaches are more meaningful than others. Since our primary goals are to assess drought impacts and drought management, data on temperature, precipitation, and drought severity are applied to time and space scales to describe climatological conditions that impact agriculture, water resources, energy production, human health, transportation, and the environment.

We use the phrase "1987-89 drought" rather loosely and refer to the drought that was most severe during the summer of 1988 over much of the U.S., but that began in many areas in 1987 and continued into 1989. At this writing, however, the drought has yet to relinquish its grip on some areas. We also note that some areas experienced almost continuous drought since 1984, and others did not exhibit drought conditions until mid-1988. In addition, we submit that multiyear drought persistence is not at all unusual and that this drought's epithet may not be written for several more years.

## The Climatological Setting

The 1987-89 drought appeared especially severe in contrast to preceding conditions. Beginning in the winter of 1984, severe or extreme drought—defined by Palmer Hydrological Drought Indices (PHDI) of ≤–3—was essentially absent from the country[1] (Figure 2.1). This is an unusual feature in itself. For the next three years, through 1986, the area of the U.S. experiencing severe or extreme drought increased to about 10% each summer, but contracted to near zero each winter. Then, during the summer of 1987, nearly 20% of the country began to experience severe or extreme drought, and during the winter of 1987-88, more than 10% of the country remained in severe or extreme drought. This first dry winter in several years set the stage for the exceptionally dry spring and summer of 1988.

## Classification and Evolution

The drought's evolution, as illustrated in Figure 2.1, can be partitioned into three phases: (1) *development*, occurring from late 1987 to July 1988; 2) *slow decay*, occurring from August 1988 to March 1989; and 3) *redevelopment*, occurring from April to July 1989.

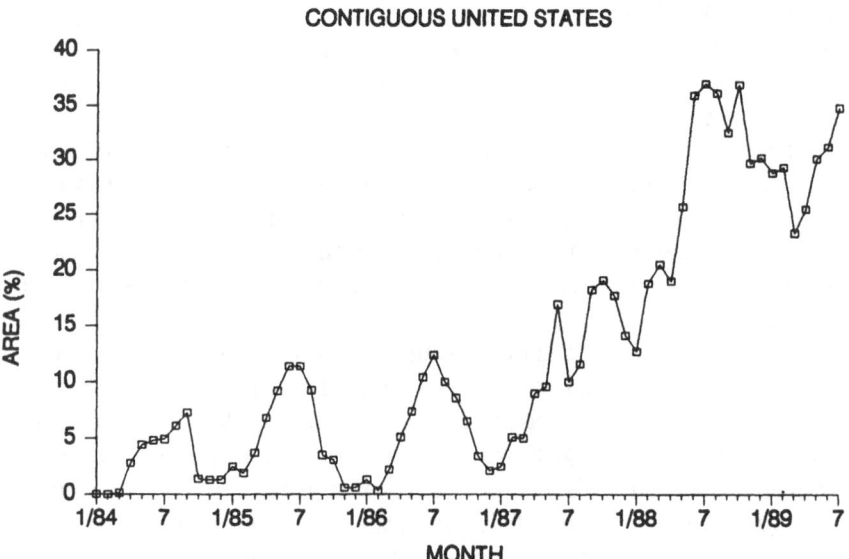

**Figure 2.1**
**Percent of the contiguous United States in severe or extreme drought (Palmer Hydrological Drought Index ≤–3) from January 1984 to July 1989**

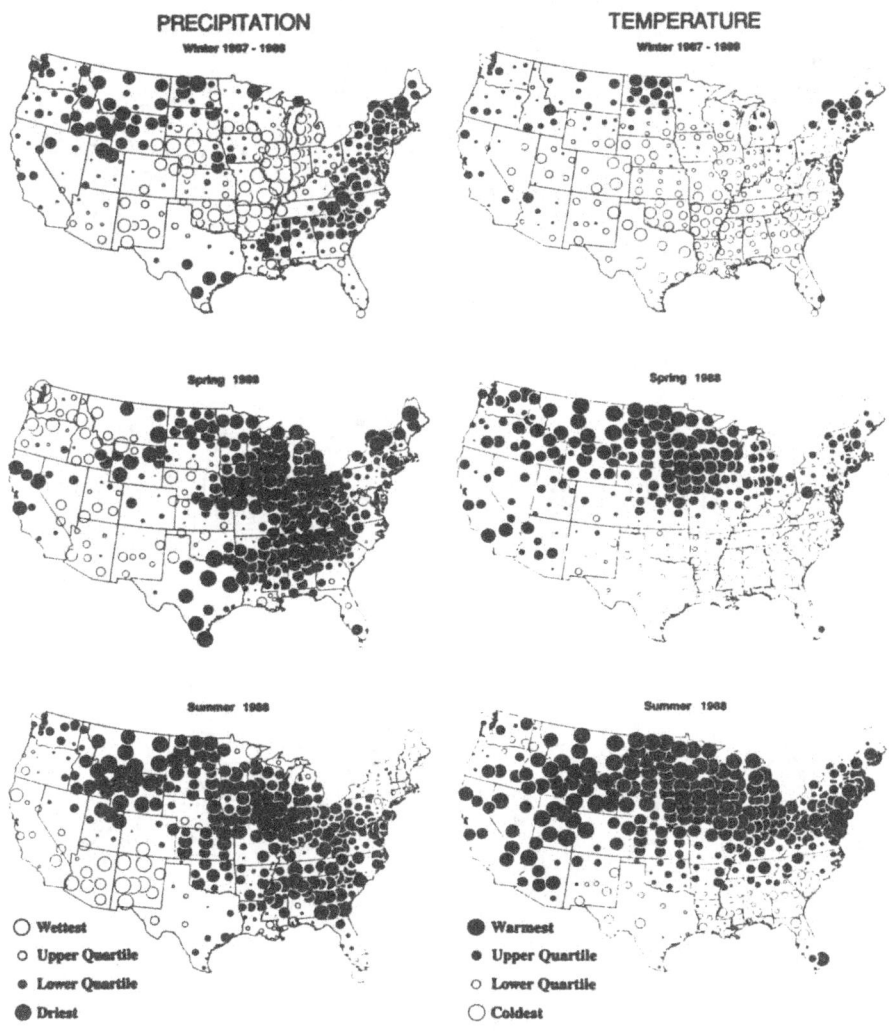

**Figure 2.2**
The severity of seasonal precipitation *(left panels)* and temperature anomalies *(right panels)* for the contiguous United States from winter 1987 to summer 1988. Both hot and dry conditions are represented by dark circles. Severity of both conditions in each of the 344 climate divisions is directly proportional to the diameter of the circles.

PALMER HYDROLOGIC DROUGHT

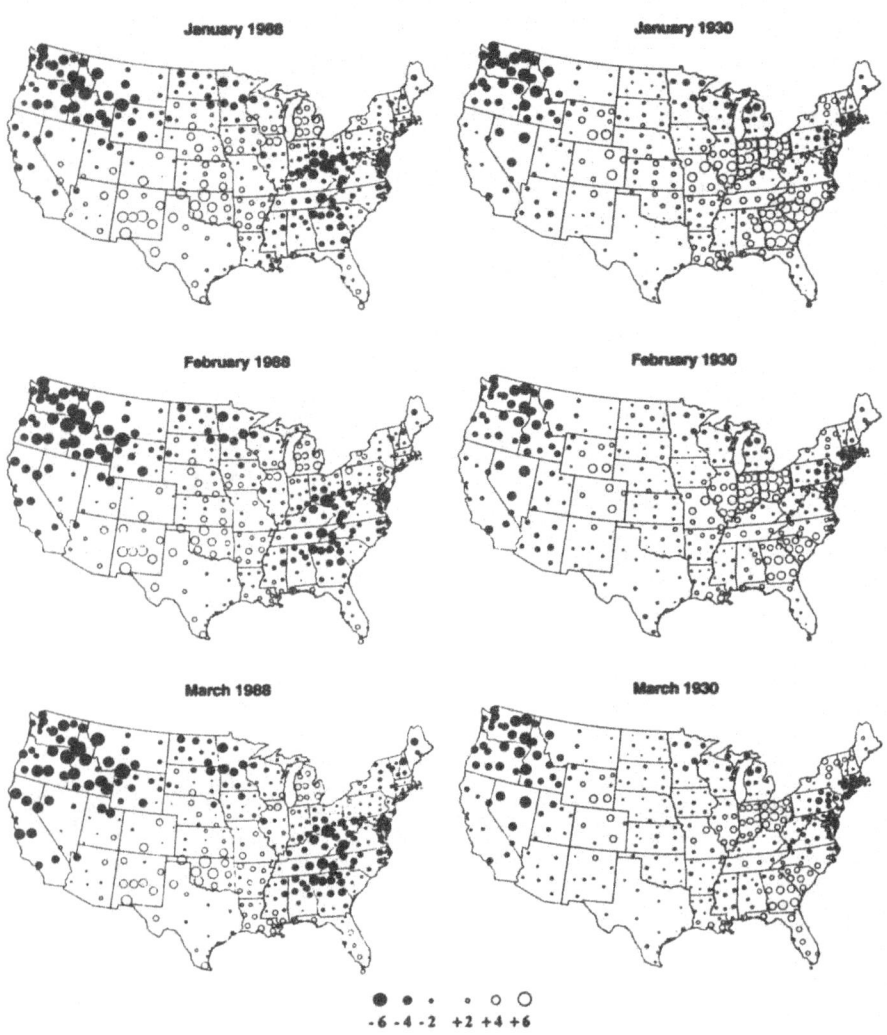

### Figure 2.3

The severity of drought and wetness during January-July 1930 and January-July 1988 as represented on this and the following page by the Palmer Hydrological Drought Index for each of the 344 climate divisions. Negative values indicating drought are denoted by dark circles and positive values indicating wet spells are represented by open circles. The magnitude of a drought or wet spell is proportional to the diameter of its circle. Values below (above) -4 (+4) reflect extreme conditions.

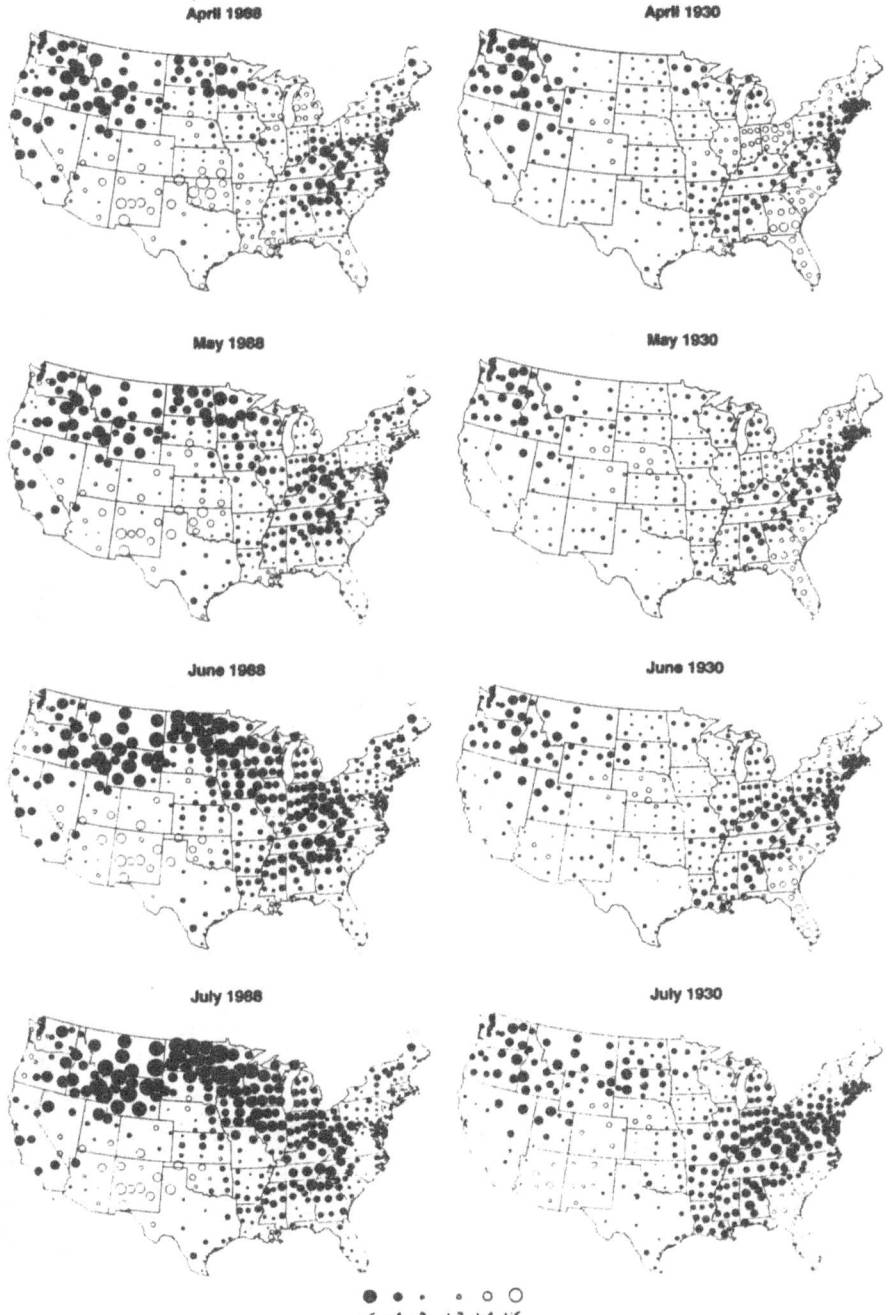

Figure 2.3 (*continued*)

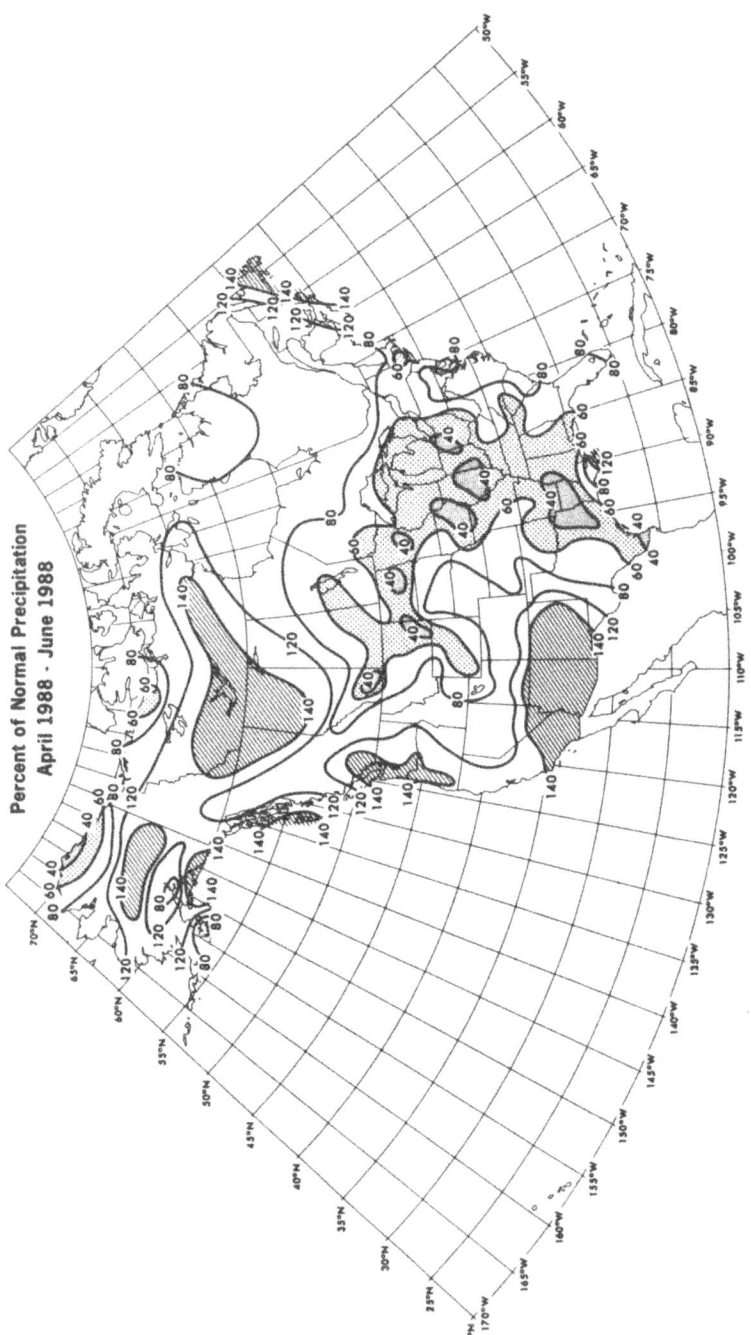

**Figure 2.4**

Percent of normal (or mean) precipitation from April through June 1988 for North America. Stippled areas represent those with less than 60% of normal precipitation, cross-hatched areas represent those with more than 140% of normal precipitation.

*Development*

During the fall and early winter of 1987, dry conditions began to develop over much of the Northwest and eastern regions of the country (Figure 2.2 on page 13). This occurred at a time when a significant portion of the annual total precipitation normally falls in these areas. During spring and summer, much of the central and north-central regions experienced extreme dryness, again at a time when the normal annual peak of precipitation is expected. In addition, these portions of the country also experienced temperatures at or above record levels, as reflected in Figure 2.2. In the past, dry conditions under these circumstances often led to rapid drought development (Karl, et al. 1987), and they did so again in 1988.

By late 1987, the Southeast had already experienced drought for several years, and at the beginning of 1988, it was joined by two other distinct areas of drought—the upper Midwest and the Pacific Northwest/Northern Rockies (Figure 2.3 on pages 14 and 15). The drought then expanded into an inverted U-shaped pattern, extending from southern California to southern Georgia. Sandwiched between this large area of drought was an area of moist conditions in the southwestern Great Plains and Southwest.

The drought extended beyond U.S. borders as the 1988 spring-summer dryness stretched into southern Canada (Figure 2.4 on page 16). From April to June, precipitation was less than 60% of normal from southern Manitoba, Saskatchewan, and Alberta to the Gulf of Mexico. Unfortunately for Canada's spring wheat belt, this is normally the wettest time of the year. During the worst of the drought, Mexico enjoyed the wetter conditions that prevailed in the American Southwest.

*Slow Decay*

During the fall of 1988 and subsequent winter, the drought decreased in severity in some areas of the United States. Particularly wet weather in the Pacific Northwest and from the Great Lakes to the Gulf Coast alleviated short-term moisture shortages in the fall (Figure 2.5 on page 19). It is worth noting that one requirement of the PHDI is the continuation of abnormally dry and/or hot weather in order to signify ongoing severe drought. Near-normal or even moderately below-normal precipitation, given enough time, will lessen or eliminate an indication of drought (Karl, et al. 1987). According to the PHDI, in several areas of the U.S., fall and winter dryness was not sufficient to maintain the midsummer intensity of the drought.

The winter of 1988-89 is best described as a period of extremes, either very wet or very dry. Dry weather returned to portions of the Great

Lakes, the West Coast, the Intermountain Region, and much of the area east of the Appalachians (Figure 2.5). By February 1989, dry conditions in the East produced mild to moderate drought in most areas, although other regions severely affected during the previous spring and summer showed significant improvement. One notable exception occurred in Kansas; although the area was only moderately affected by the drought during spring and summer of 1988, two back-to-back very dry seasons, fall and winter, were enough to put many portions of the state in severe to extreme drought.

## Redevelopment and Southwestward Expansion

During the spring of 1989, heavy precipitation in the East and the Pacific Northwest alleviated drought conditions, but dry conditions persisted in the central Great Plains, much of the Midwest, and the Southwest (Figure 2.6 on page 20). Many above-normal temperatures were coupled with the dry conditions. For the first two months of the summer of 1989, dry and very hot weather continued in the Southwest and returned to much of the northern Plains (Figure 2.7 on page 21), although much of the South and Southeast was very wet.

The PHDI responded to the wet conditions in the South and Southeast, as moderate to extremely wet conditions replaced areas of drought. Dry weather in the Southwest and northern Plains however, led to redevelopment of the drought in the northern Plains and its expansion and intensification in the southwest United States (Figure 2.8 on page 22).

## Comparisons with Past Droughts

In order to place the 1987-89 drought in climatological perspective, a comparison of these three periods of the drought should be made to previous events. During severe and extreme droughts, the most appropriate analog to an ongoing drought is often an issue of concern. Although, there are many ways to assess similarity between droughts, we chose to search for *patterns* and *severity* of past droughts that most closely matched those of 1987-89. These analogs were selected by calculating the area-weighted mean squared differences of the PHDI between 1988 and other years for each of the 344 climate divisions in the contiguous United States on a month-by-month basis from January through July.[2] The periods with the smallest sum of squared differences were cited as the best analogs.

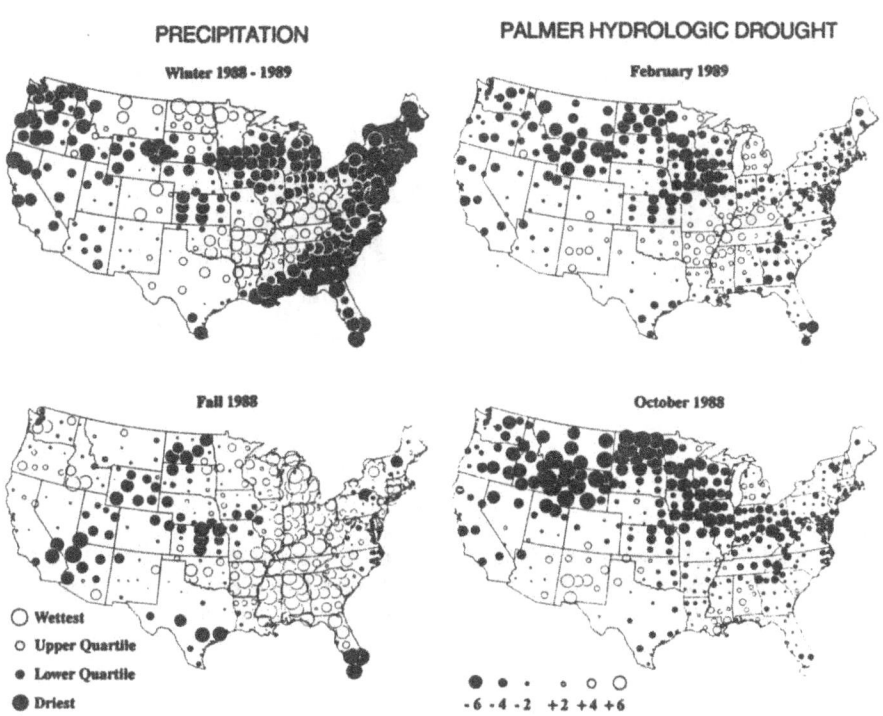

**Figure 2.5**
The severity of seasonal precipitation anomalies during fall 1988 and
winter 1988-89 *(left panel)* are reflected by the diameter of the circles;
dry conditions are represented by dark circles and wet conditions by
open circles (see Figure 2.2). The severity of drought and wetness during
October 1988 and February 1989 *(right panel)* are represented by the
PHDI for each of the 344 climate divisions; dark circles denote drought
and open circles indicate wet spells (see Figure 2.3).

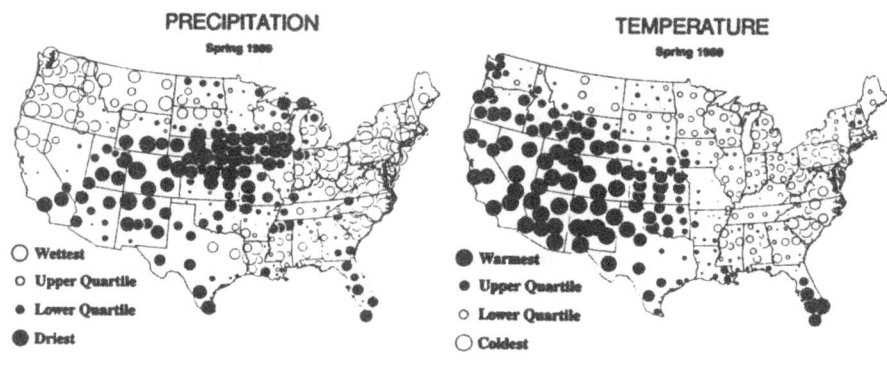

**Figure 2.6**
The severity of seasonal precipitation and temperature anomalies for
Spring 1989 are reflected by the diameter of the circles. Both hot and
dry conditions are represented by dark circles (see Figure 2.2).

*Development*

Using the PHDI, the best analog we could find for the 1988 development phase of the drought occurred during 1930 (Figure 2.3 on page 14). Drought in both years affected most of the country, except for the Southwest. Although the indices of the 1930 drought showed a number of similarities with the 1987-89 drought, the year most closely matching the pattern of actual *precipitation anomalies* during the development phase of the drought was 1931 (Figure 2.9 on pages 24 and 25). In that respect, it is a better analog than 1930, but because the area of extreme and severe PHDI by January of 1931 (Figure 2.10 on page 26) was more than twice as large as that of 1988 (Figure 2.1 on page 12), it was not selected as a good PHDI analog. The pattern of precipitation anomalies during 1931 was remarkably similar to those of 1988, and these similarities suggest that the observed patterns of dryness are a "characteristic mode" of atmospheric dryness in the U.S., that is, a pattern that shows up repeatedly. Such a pattern of dryness may be initiated by characteristic anomalies of sea-surface temperatures and subtropical convection, as suggested by Trenberth, et al. (1988).

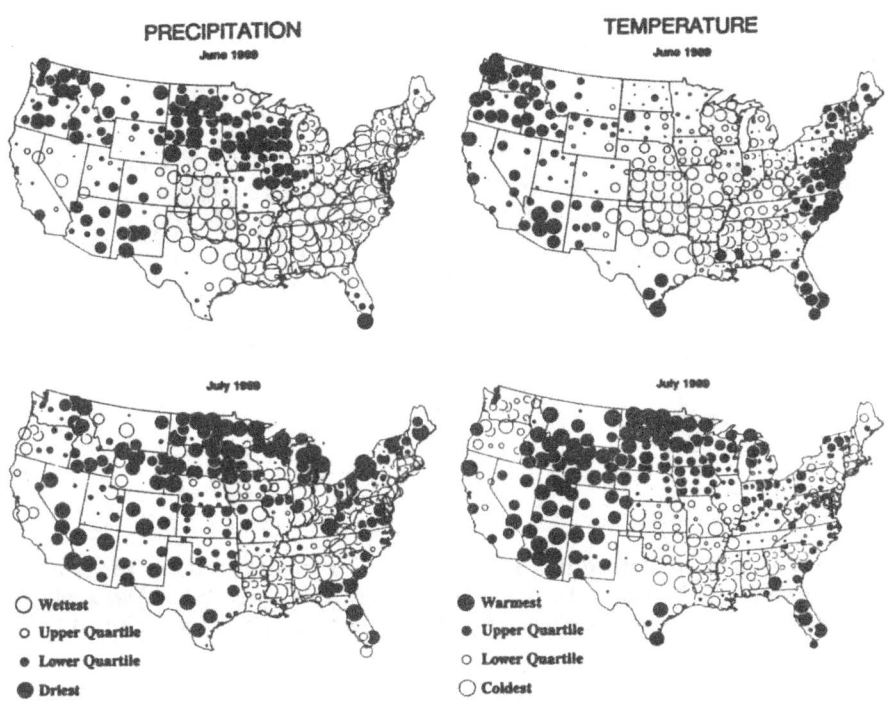

**Figure 2.7**
The severity of precipitation and temperature anomalies for June and
July 1989 are reflected by the diameter of the circles. Both hot and dry
conditions are represented by dark circles (see Figure 2.2).

PALMER HYDROLOGIC DROUGHT

Figure 2.8
The severity of drought and wetness as represented by the Palmer
Hydrological Drought Indices for March and July 1989. Dark circles
denote drought and open circles indicate wet spells (see Figure 2.3).

Actually, the national precipitation total for January through July
of 1931 was not quite as low as that observed in 1988 (Figure 2.11 on
page 26), an effect partly due to the timing of very dry conditions during
these two years. January 1931 was significantly drier than January 1988,
but June was significantly drier in 1988 than in 1931. Since June
typically has more precipitation than January on a national basis, the
year with the drier June would tend to yield lower national area-
average total precipitation. This is not simply a statistical curiosity,
because June precipitation is critical to agriculture in much of the
country, and comparisons that weigh it evenly with other months may
not reflect the social significance of drought.

As indicated in Figure 2.1 on page 12, the 1987-89 drought reached
its peak in July of 1988. Comparison of the area covered by severe and
extreme drought in 1988 with past drought years indicates that, although the drought was an historic event, it was not unprecedented
(Figure 2.12 on page 27). Indeed, by this measure, it was slightly less
widespread than the midwestern and western drought that evoked a
national emergency in the mid-1970s, and was certainly less widespread

than the well-known drought years of the 1930s and 1950s. The greatest areal extent and intensity of drought in the instrumented climate record of the U.S. was in July 1934 (Figure 2.13 on page 27). The 1934 drought is quite impressive, even when compared to the intensity and areal extent of the 1988 drought in July (Figure 2.3 on page 14).

Another perspective of the 1987-89 drought can be obtained by calculating a national PHDI using area-weighted monthly total precipitation and mean temperature values for the country as a whole (Figure 2.14 on page 28). Such calculations indicate that the value of the PHDI during 1988 has been exceeded many times in the past. The accumulated precipitation deficit during the most notable droughts of this century have also been substantially greater, especially in the 1930s and 1950s, and even during the drought of the 1960s. However, the 1960s drought was centered in the Northeast, where precipitation is relatively high and the cumulative departure from normal quickly reached large values, compared to lower, but more indicative of dry conditions, values that would have resulted from a drought in drier parts of the country.

Figures 2.12 (on page 27) and 2.14 (on page 28) reveal an important characteristic of U.S. droughts. Major drought years, such as those in the 1930s, 1950s, and the mid-1960s, were not isolated events. At least four distinct drought episodes can be discerned in the 1930s: 1930-31, 1934, 1936, and 1939-40. Similarly, several "waves" of drought amelioration and reintensification occurred during the 1950s and 1960s. These ensembles of dry seasons or years suggest that drought-producing climate patterns may have a "memory" lasting more than a few seasons or years (Karl 1988). Some notable droughts, however, have been more isolated events amid wet spells: the 1976-77 drought in the West and Midwest and the 1980 heat wave and drought over the central U.S., for instance. Thus far, the dry spell of 1987-89 qualifies more as an isolated event, though, of course, its end or persistence cannot be predicted with any certainty.

*Slow Decay*

The decrease in areal extent of drought from August 1988 to March 1989 was quite comparable to that observed during oscillations of the 1930s drought episode. Figure 2.15 depicts changes in areal extent of the 1936 drought, most notably from an August peak of about 45% of the United States in severe or extreme drought, to about 20% by March of 1937. This decrease was somewhat greater than that observed in 1988 (Figure 2.1 on page 12).

PRECIPITATION

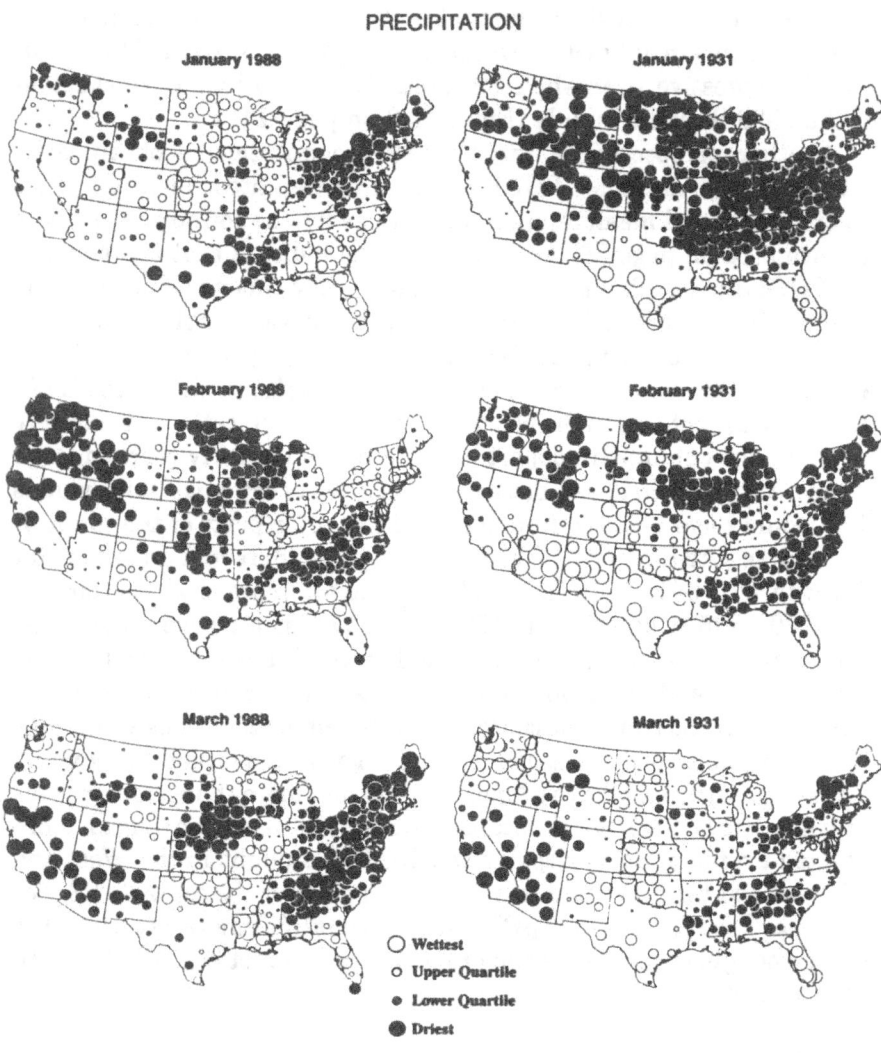

**Figure 2.9**
The severity of monthly precipitation anomalies for January-July 1988
*(left panel)* and January-July 1931 *(right panel)* are reflected on this and
the following page by the diameter of the circles. Dry conditions are
represented by dark circles and wet conditions by open circles (see Figure
2.2).

PRECIPITATION

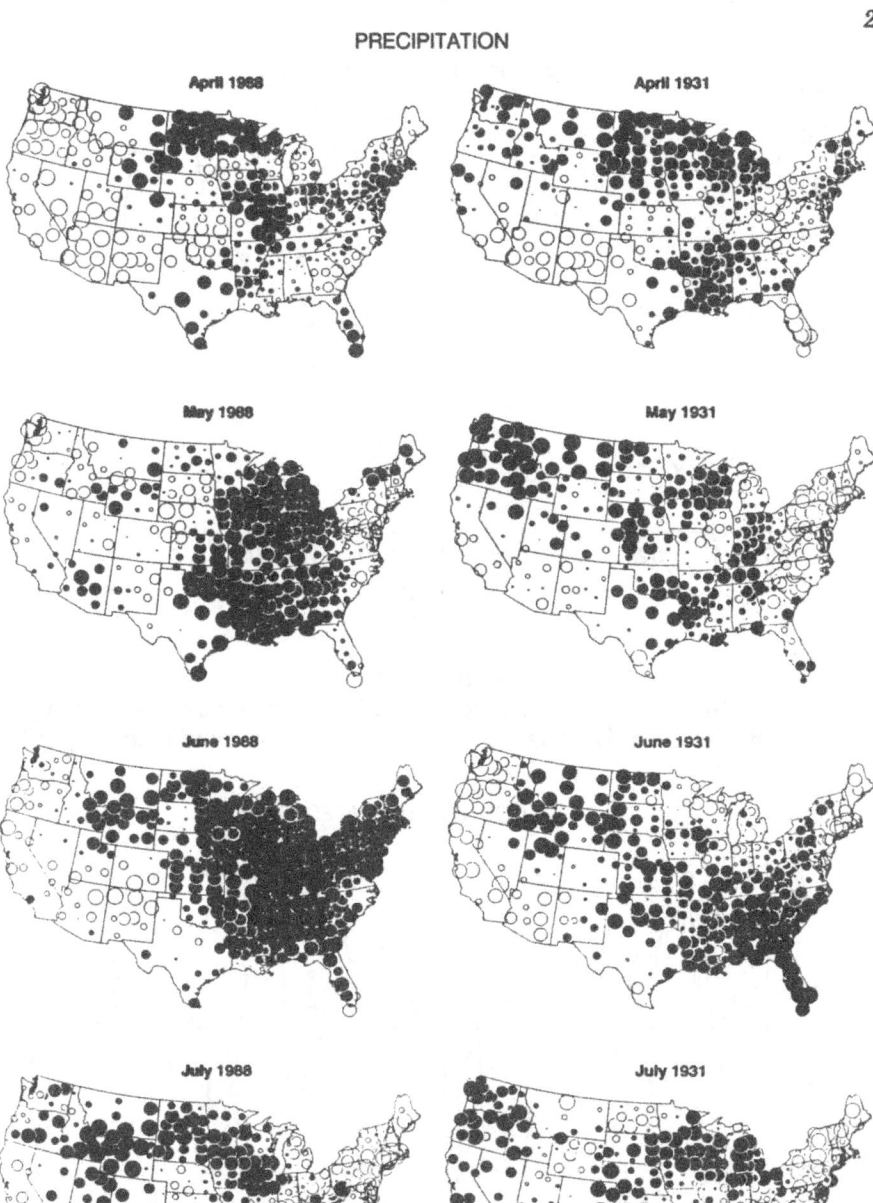

Figure 2.9 *(continued)*

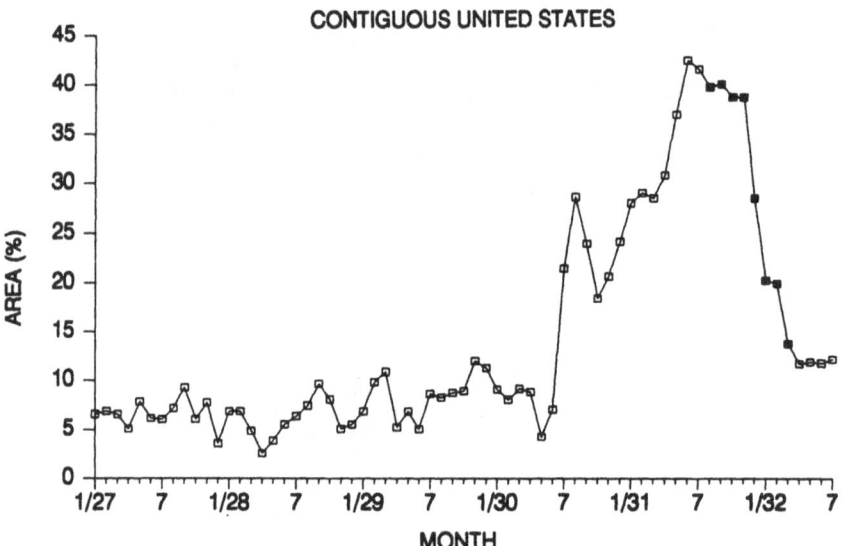

**Figure 2.10**
**Percent of the contiguous United States in severe or extreme drought**
**(Palmer Hydrological Drought Index ≤-3) from January 1927 to July 1932**

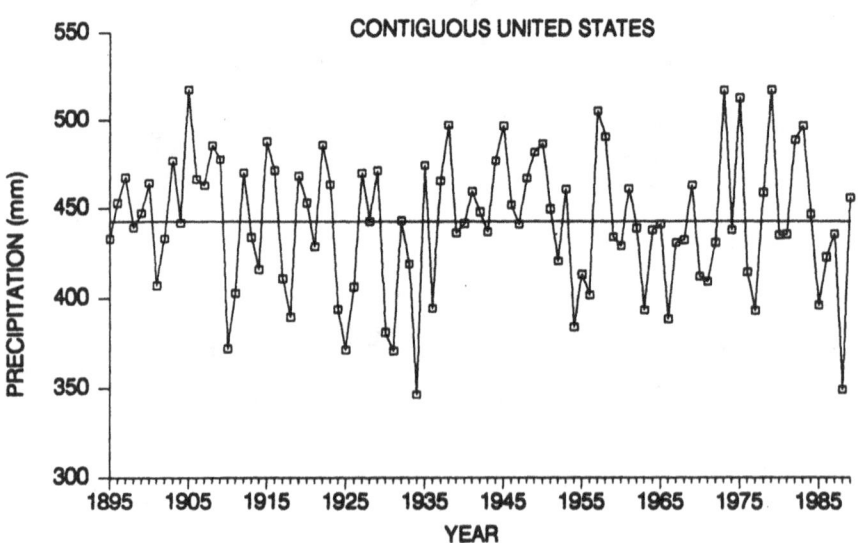

**Figure 2.11**
**Total precipitation occurring from January through July for years**
**1895-1989, area-weighted over the contiguous United States**

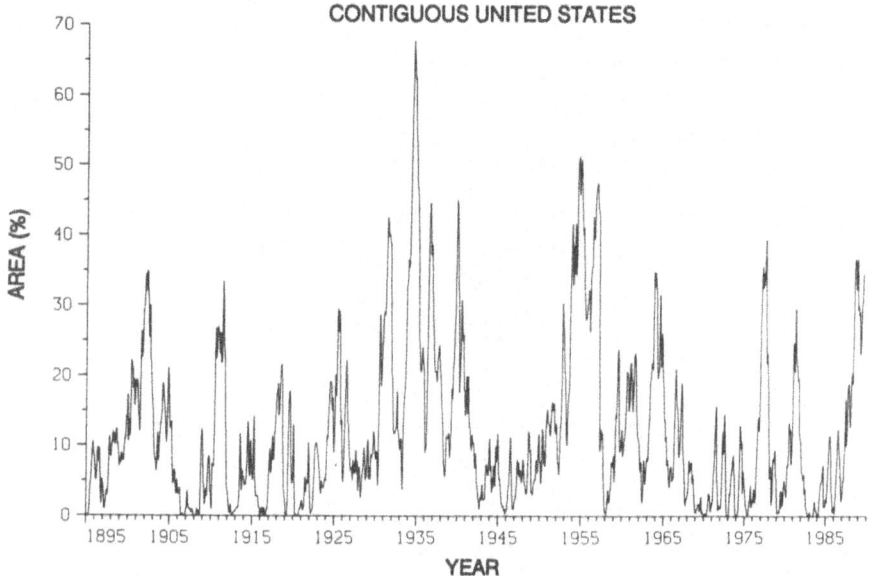

**Figure 2.12**
Percent of the contiguous United States in severe or extreme drought
(Palmer Hydrological Drought Index ≤-3) from 1895-1989

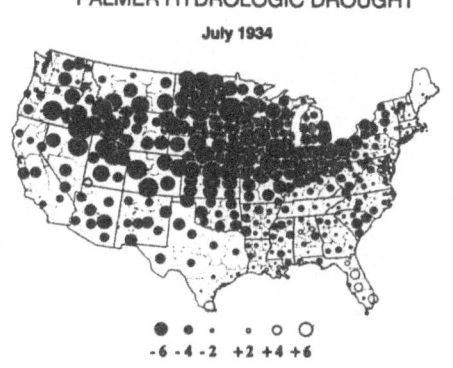

**Figure 2.13**
The severity of drought and wetness during July 1934 as represented
by the Palmer Hydrological Drought Index for each of the 344 climate
divisions. Dark circles denote drought and open circles indicate wet spells
(see Figure 2.3).

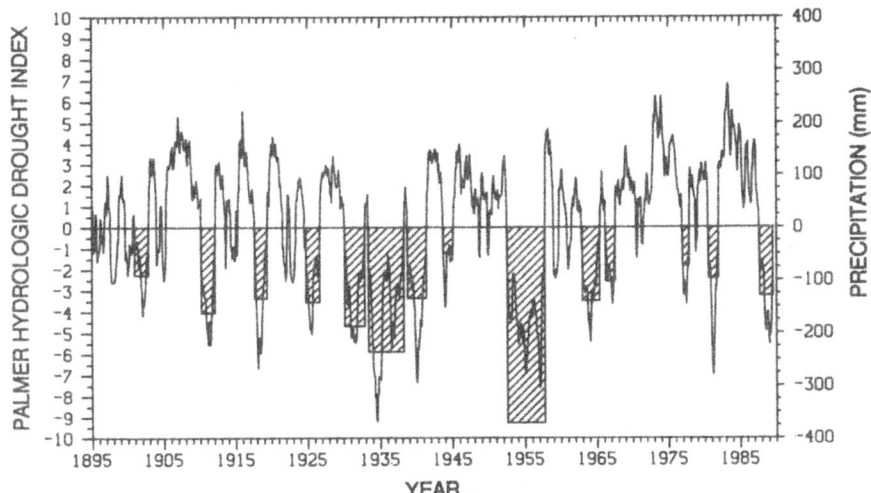

**Figure 2.14**
Palmer Hydrological Drought Indices (PHDIs) calculated using area-weighted total precipitation and temperature for the contiguous United States. Accumulated national average total precipitation deficiencies during major droughts (PHDI ≤-3) are given by the hatched rectangles using the right-hand scale. The length of major droughts is depicted by the width of the rectangles along the time axis.

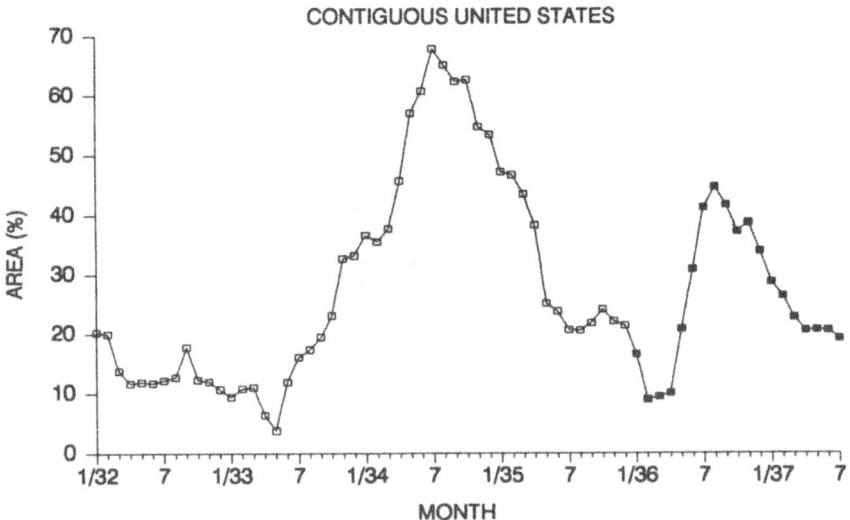

**Figure 2.15**
Percent of the contiguous United States in severe or extreme drought (Palmer Hydrological Drought Index ≤ -3) from January 1932 to July 1937

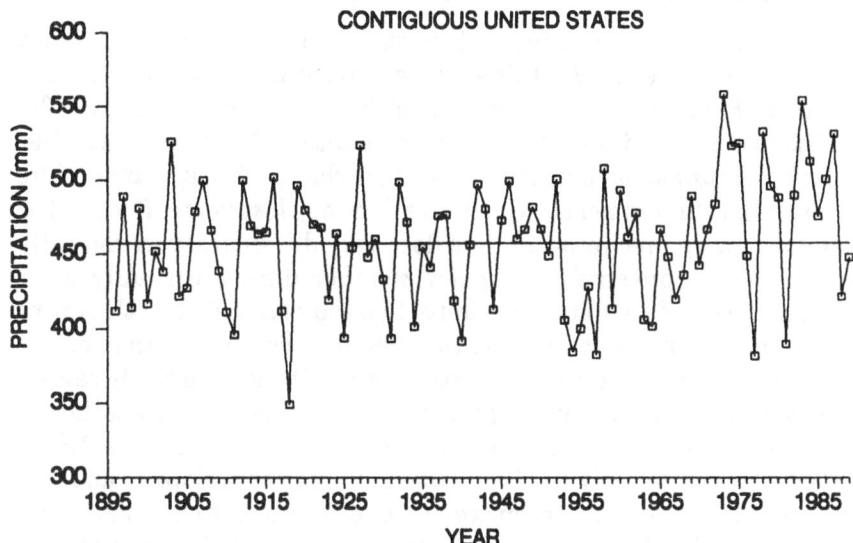

**Figure 2.16**
**Total precipitation for August-March from 1895-96 to 1988-89, area-weighted over the contiguous United States**

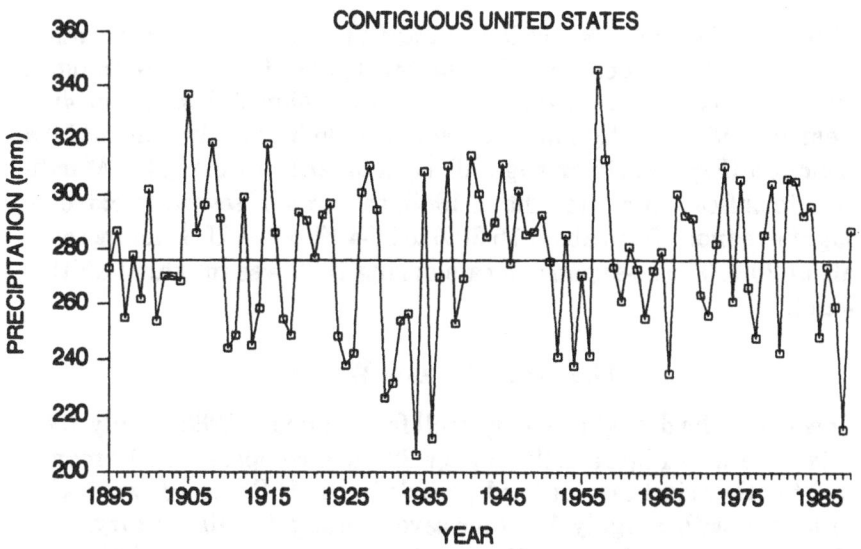

**Figure 2.17**
**Total precipitation for April-July from 1895 to 1989, area-weighted over the contiguous United States**

*Redevelopment and Southwestward Expansion*

Total precipitation over the contiguous United States during the slow decay phase of the 1987-89 drought was near the long-term normal (Figure 2.16 on page 29), again comparable to that observed in 1936-37. It should be noted that the timing and location of wet or dry weather affects precipitation anomalies and how they change through time. Little or no precipitation in normally dry locations has less of an affect on the total nationwide precipitation than does very dry weather in normally wet locations. Conversely, very dry weather during normally drier months has less of an anomalous impact on the national total than very dry weather during wet months. In 1988, dry weather in the central and southwestern U.S. during fall and winter, although relatively severe, was somewhat offset in terms of national anomalies by wet weather in some of the wetter portions of the country (Figure 2.5 on page 19).

From March to July of 1989, the drought redeveloped, and the proportion of the country in severe or extreme drought expanded from 25% to about 35%. Because this redevelopment included expansion into the normally drier parts of the country, such as the Southwest (Figure 2.8 on page 22), and because the normally wet East was excessively wet during the spring and summer months (Figures 2.6, on page 20, and 2.7, on page 21), the total national precipitation for the redevelopment phase was actually *above* the long-term normal (Figure 2.17 on page 29).

This second emergence was less marked than similar patterns of other severe drought episodes. In 1934, the third year of the 1930s drought, total area in severe or extreme drought redeveloped from about 40% to near 70% of the contiguous U.S., and in 1936 it grew from about 10% to near 45% (Figure 2.15 on page 28). By comparison, during the March-July intensification of drought in 1988, the area of severe to extreme drought expanded from about 15-20% to 35-40% of the U.S., an increase of about 20%. The re-expansion rate during 1989 was roughly half this amount.

### The Drought as a Whole

If we view the drought as a dry spell from January 1988 to July 1989 (the last point in Figure 2.18 on page 32), and compare this 19-month period to other similarly defined periods in the U.S. record, we find it is in league with roughly 10 other severe droughts this century.

Looking just at calendar 1988, we find that it was a remarkably dry year. Only three years—1910, 1917, and 1956—had lower annual precipitation totals, with 1963 not far behind (Figure 2.19 on page 32). Another way to view the 1988 precipitation pattern is to standardize the departures from normal precipitation in each climate division before summing them to get the national total—effectively neutralizing large anomalies that occur in normally wet areas of the country. The *gamma* distribution, a statistical dispersion that can be skewed toward large values in a manner similar to actual precipitation data, was used for this purpose. These standardized precipitation anomalies again show that 1988 is not unprecedented, but compares with roughly 10 other years this century (Figure 2.20 on page 33). Note also that the 1987-89 drought followed a period of consistently wet weather.

The overall pattern of the 1987-89 drought most resembles the drought of 1936. We searched the U.S. climate record for the 19-month period that most closely matched the 1988 drought development, slow decay, and redevelopment according to total area of the U.S. in severe or extreme drought (Figure 2.1 on page 12). This was accomplished by calculating the area under the curve in Figure 2.1 month-by-month over the 19-month period from January to July 1989 and summing the squared difference of this value with previous years for the same period. The period of drought from January 1936 to July of 1937 was the best match. The 1936 drought developed rapidly in the spring and early summer and then began a slow decay in the late summer of 1936 (Figure 2.15 on page 28). The redevelopment phase, however, is missing from the 1936 drought.

## Regional Patterns of the 1987-89 Drought

Thus far, we have seen that the 1987-89 drought, although quite intense in its own right, was equaled or exceeded by several previous droughts during the 20th century. It did, however, exhibit several geographic and temporal patterns that are particularly noteworthy for their biophysical and social impacts.

For example, several critical agricultural regions and watersheds experienced extraordinarily dry weather during the drought (Figure 2.21 on page 34). During April-June of 1988, the period of most rapid intensification, total precipitation over the country's primary corn and soybean growing areas was the lowest of the twentieth century—drier than the well-known droughts in 1934 and 1936 (Figure 2.22 on page 35). This pattern had enormous economic consequences, as described in the next chapter. Even though precipitation in these areas improved substantially during the same period in 1989, levels were still below normal. The proportion of corn and soybean growing areas with severe or extreme drought during 1988 was over 70%, which was exceeded only in the 1930s and equaled only in 1954 (Figure 2.22).

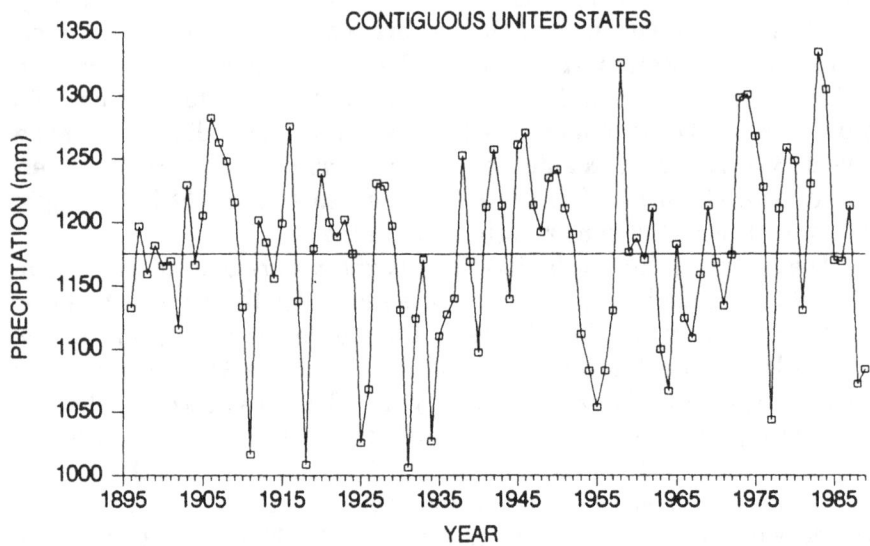

**Figure 2.18**
Total precipitation that occurred from 1895-96 to 1988-89 during 19-month periods that began in January and ended in July of the following year, plotted for the year in which the 19-month period ends

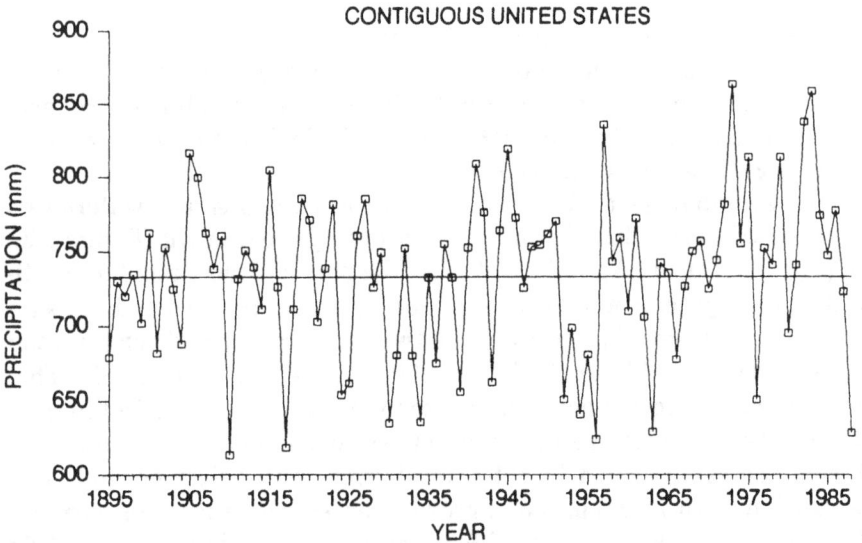

**Figure 2.19**
Total annual precipitation from 1895 to 1989, area-weighted over the contiguous United States

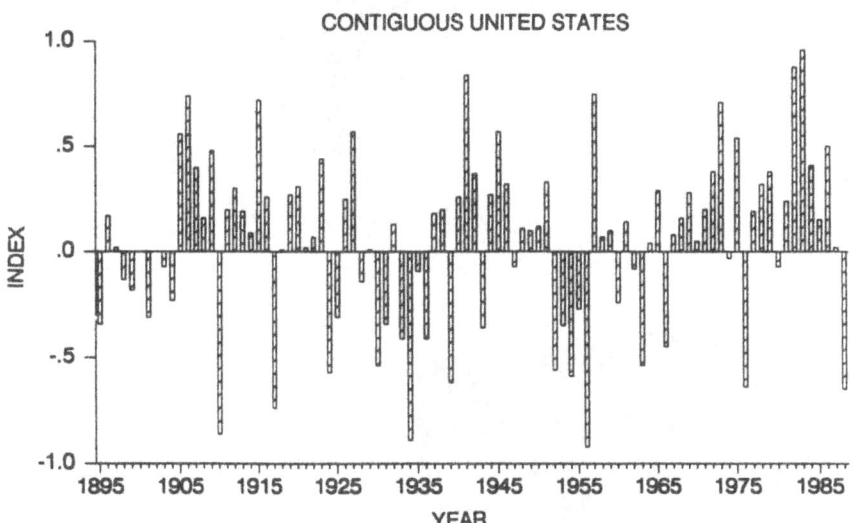

**Figure 2.20**
Index of total annual precipitation, area-weighted over the contiguous
United States from 1895-1988. The index places the annual precipitation
anomalies for each climate division on the same standard statistical
distribution (the *gamma* distribution) and uses the annual "z-scores" from
this statistical distribution.

While these record-low rainfalls occurred, the total precipitation that
fell over the country's largest watershed, the upper Mississippi River
drainage area, was also near record-low levels—only 1934 and 1936 were
drier (Figure 2.23 on page 36). The result, of course, was unprecedented
low river levels and navigation problems on the Mississippi River, as
described in Chapter 4.

The lowest growing-season (April-July) precipitation since 1936 in
the nation's spring wheat belt occurred in 1988 (Figure 2.24 on page
37), and dry conditions resurfaced during the drought's redevelopment
in 1989. Likewise, the proportion of the area in extreme or severe drought
during 1988 was exceeded only in 1934. In 1989, however, well over half
the area was still in extreme or severe drought, but this was by no means
unprecedented.

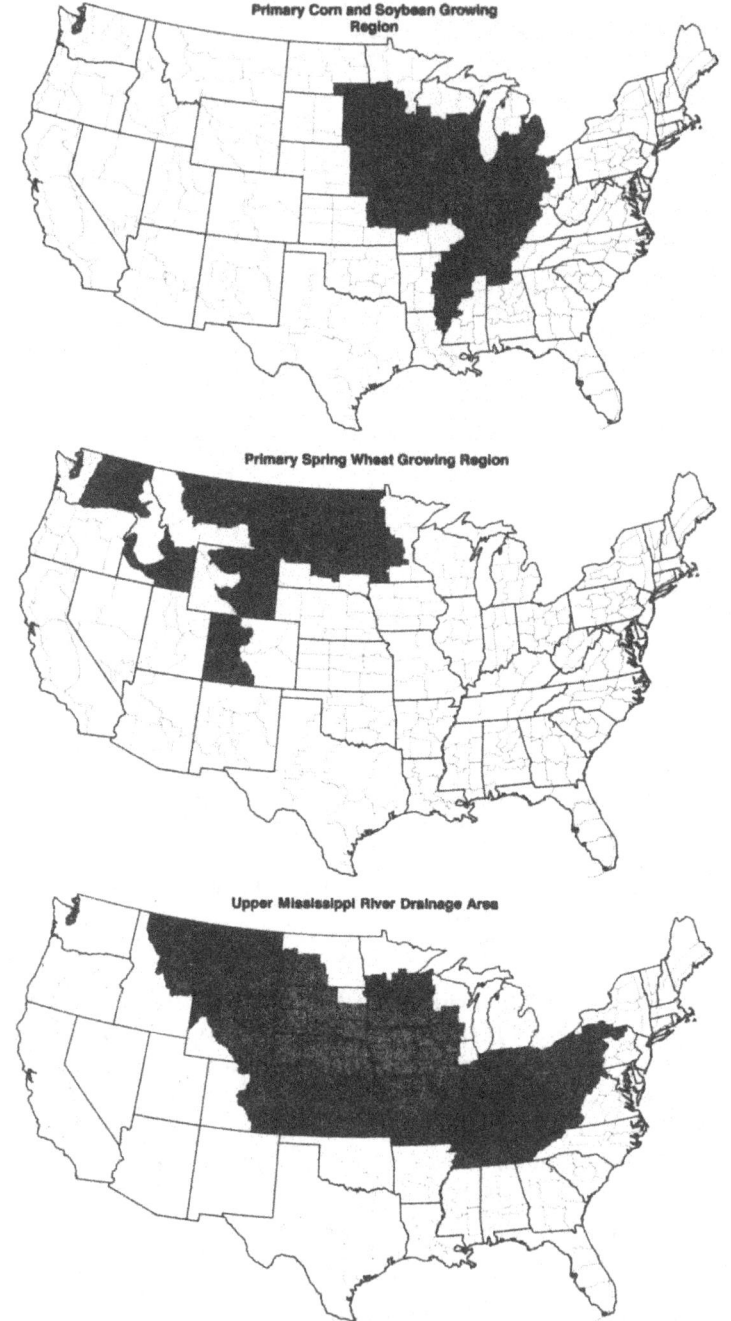

**Figure 2.21**
Each shaded area defines a region of agricultural or hydrological importance

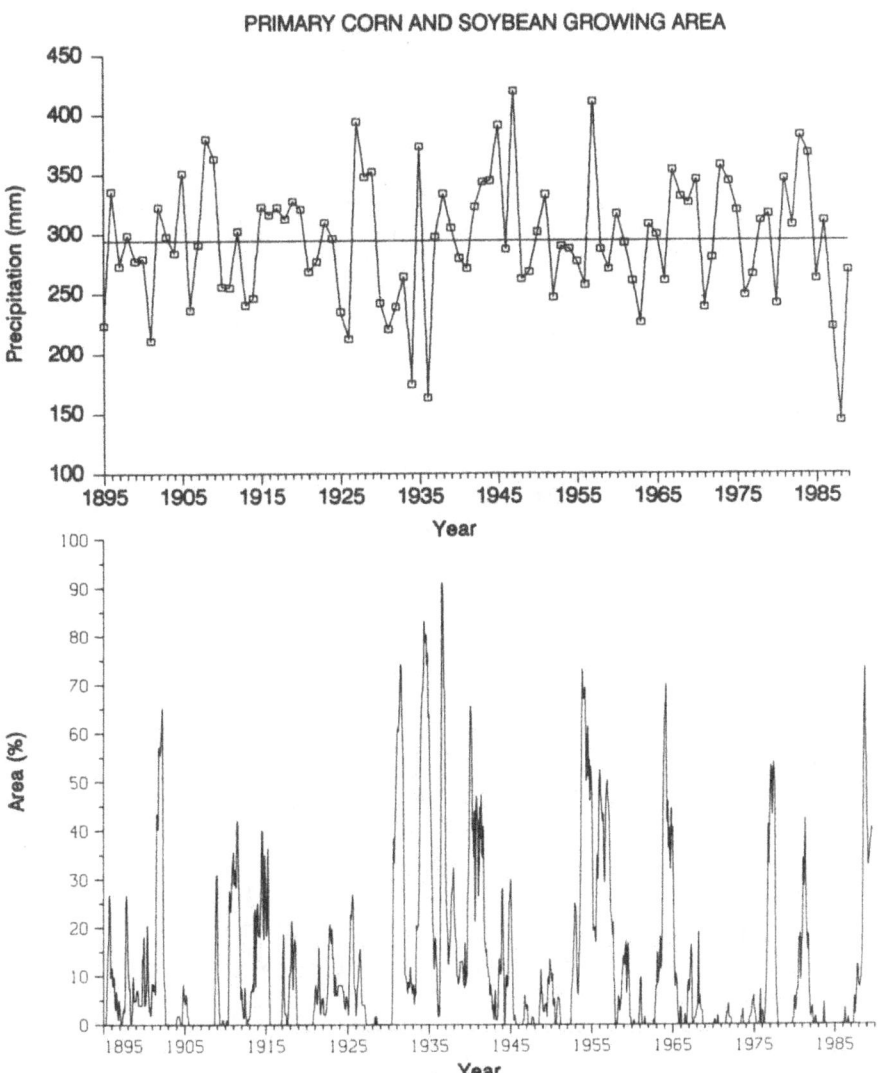

**Figure 2.22**
Total precipitation (*top*) for April-June from 1895 to 1989, area-weighted
over the primary corn and soybean belt of the contiguous United States
(see Figure 2.21), and the percent of this area in severe or extreme drought
(*bottom*)

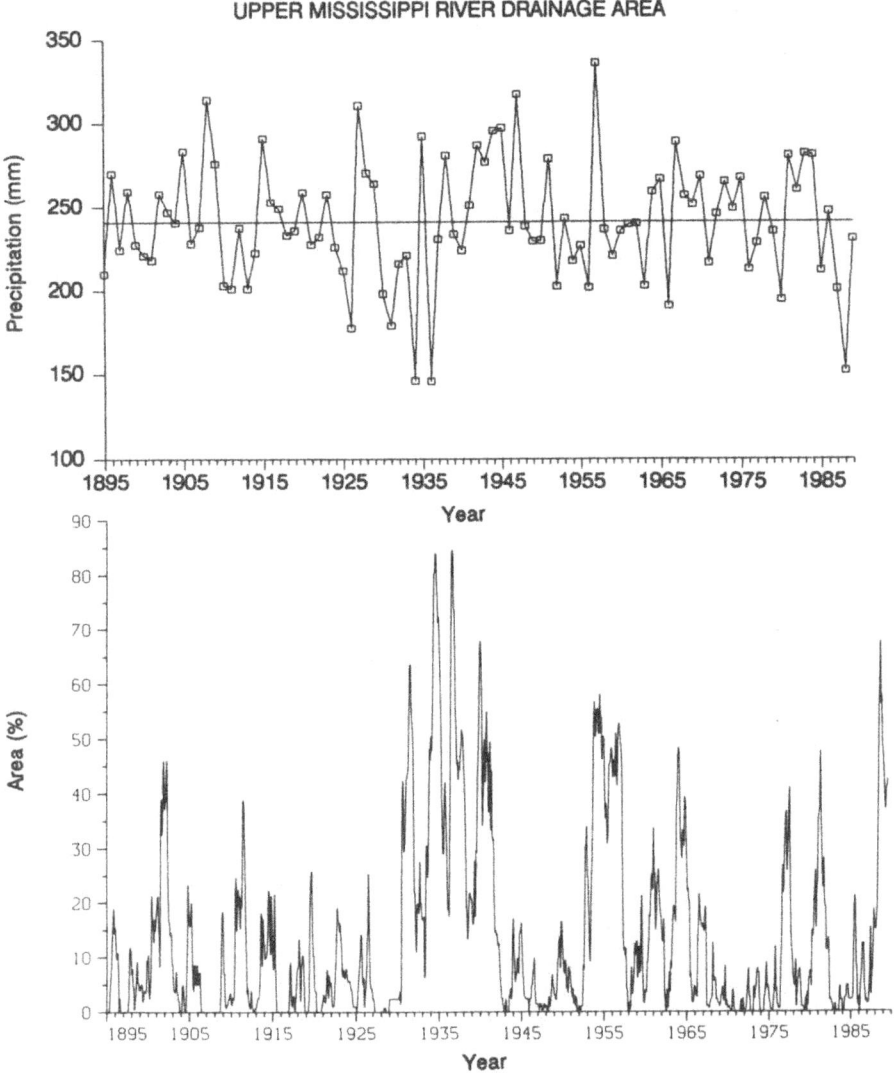

**Figure 2.23**
Total precipitation (*top*) for April-June from 1895 to 1989, area-weighted
over the Upper Mississippi River Drainage Basin (see Figure 2.21), and
the percent of this area in severe or extreme drought (*bottom*)

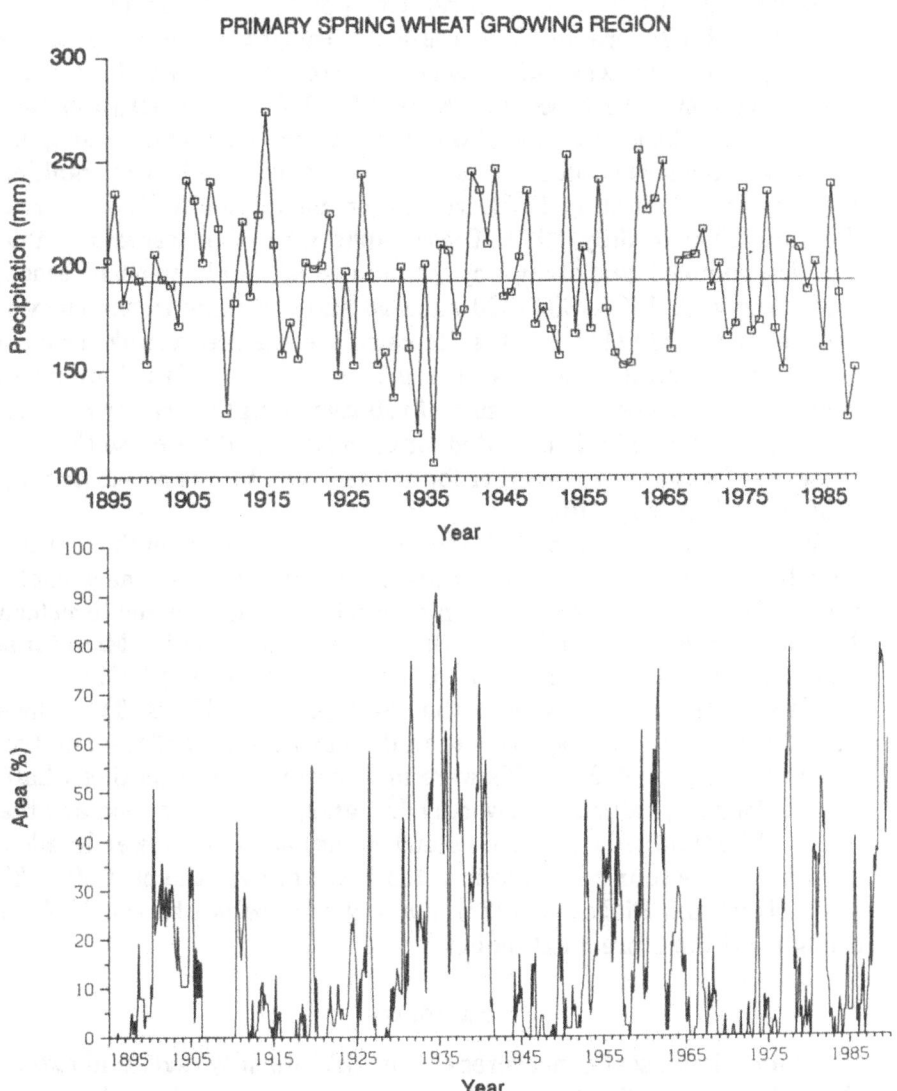

**Figure 2.24**
Total precipitation (*top*) for April–July from 1895 to 1989, area-weighted
over the Spring Wheat Belt (see Figure 2.21), and the percent of this area
in severe or extreme drought (*bottom*)

## The Decadal Pattern of 20th Century Droughts

Perhaps the most remarkable aspect of the 1987-89 drought was that it occurred after two decades of relatively wet conditions nationwide. This places the severe consequences of the drought in context, and may help explain why it was such a shock to the country's natural resources management systems. Since the decade of the 1980s is virtually complete (less five months at the time of this writing), we were able to compare the nine calendar decades of this century. We did this by first ranking the severity of monthly PHDI values for each climate division from January 1895 to July 1989. These ranks were then separated into decades, summed, and divided by the number of months in each decade, beginning with 1900-1909 (120 months) and ending with the current decade, 1980-1989 (115 months), to calculate the mean rank for each decade. The resultant decade values, after subtraction of the total mean rank for all decades, is a measure of relative drought severity for each decade. This value is then plotted for each climate division so that the patterns of wet and dry weather for a decade can be compared visually (Figure 2.25 on page 40).

What we find is that the 1970s were the wettest decade of the century, with the 1940s and 1980s close runners-up. For the decade as a whole, the 1980s were consistently dry in only a few areas: the northwestern Plains and much of Wyoming and the southern Appalachian Mountains through the mid-Atlantic seaboard (see Karl and Young 1987).

This contrasts sharply with the 1930s and the 1950s. The 1930s droughts tended to be most severe in the northern half of the country, and the droughts of the 1950s were most severe in the southern half. The 1960s also produced noteworthy droughts, both in the East and the West, while the 1970s and 1980s stand out as back-to-back wet decades. In general, the country has been getting wetter over the past 20 to 30 years (Karl and Riebsame 1984), making dry spells like the 1987-89 drought all the more outstanding.

## Summary

At a national scale, measures of drought intensity and areal extent show that the 1987-89 drought was one of roughly 10 of the worst drought-years this century. It was not the worst drought in history, nor was its intensity in any given season or month unprecedented in the last 100 years, despite its extraordinary impacts and the widespread public concern it evoked. However, 1988 was the fourth driest year of this century for the country as a whole, and in some regions such as the Midwest and northern Great Plains, the drought was indeed a rare event, matching or exceeding the drought conditions of the 1930s Dust Bowl.

While the 1987-89 drought was not unprecedented, it stands in sharp contrast to the unusually wet climate conditions over much of the nation during the previous two decades. Through 1987, the decade of the 1980s was actually notable for excess rainfall and flooding that produced, for example, record high water levels in the Great Salt Lake and the Great Lakes (Karl and Young 1986; Changnon 1987). Of course, the nation experienced the usual array of significant, but brief, regional droughts in the 1980s: in the Midwest (Illinois Agricultural Statistics Service 1983), the southern Appalachian Mountains (Karl and Young 1987), and southern California. The summer of 1980 was also exceptionally hot and dry over the central U.S. (Karl and Quayle 1981). Yet, in broad climatological terms, the 1970s and 1980s will be remembered as relatively wet over the country, a pattern that insured the severe dryness in 1988 received widespread public attention, and perhaps heightened its impacts on resources systems accustomed to more plentiful precipitation. As the first large-scale, persistent dry spell in roughly 30 years, the 1987-89 drought affected many resources managers and planners who had never experienced major drought before.

### Notes

1.  The Palmer Hydrological Drought Index (PHDI) is a measure of the long-term (deep soil) moisture supply at a given location. PHDI values typically span a range from -6.0 to +6.0, with values of -3.0 to -4.0 indicating severe drought that begins to affect most water resources systems, and those less than -4.0 representing the most intense droughts that affect most natural systems adversely. (Positive values in these ranges indicate severe or intense wetness, often leading to flooding.) The Palmer Drought Severity Index (PDSI) is a related, and frequently used, index more responsive to short-term climate conditions. For example, the PDSI moves quickly from drought (negative values) to normal (near-zero values) the first month that *begins* a string of months with sufficient moisture to replenish topsoil moisture and meet evapotranspiration demand. In comparison, the PHDI does not return to normal until *all* the required moisture has been received, and thus it better reflects the cumulative nature of drought development and impact on natural systems.

2.  Each state has been subdivided into a number of smaller climate divisions. The number of divisions ranges from one for small states such as Rhode Island to 10 for states such as Texas.

DROUGHT / WETNESS

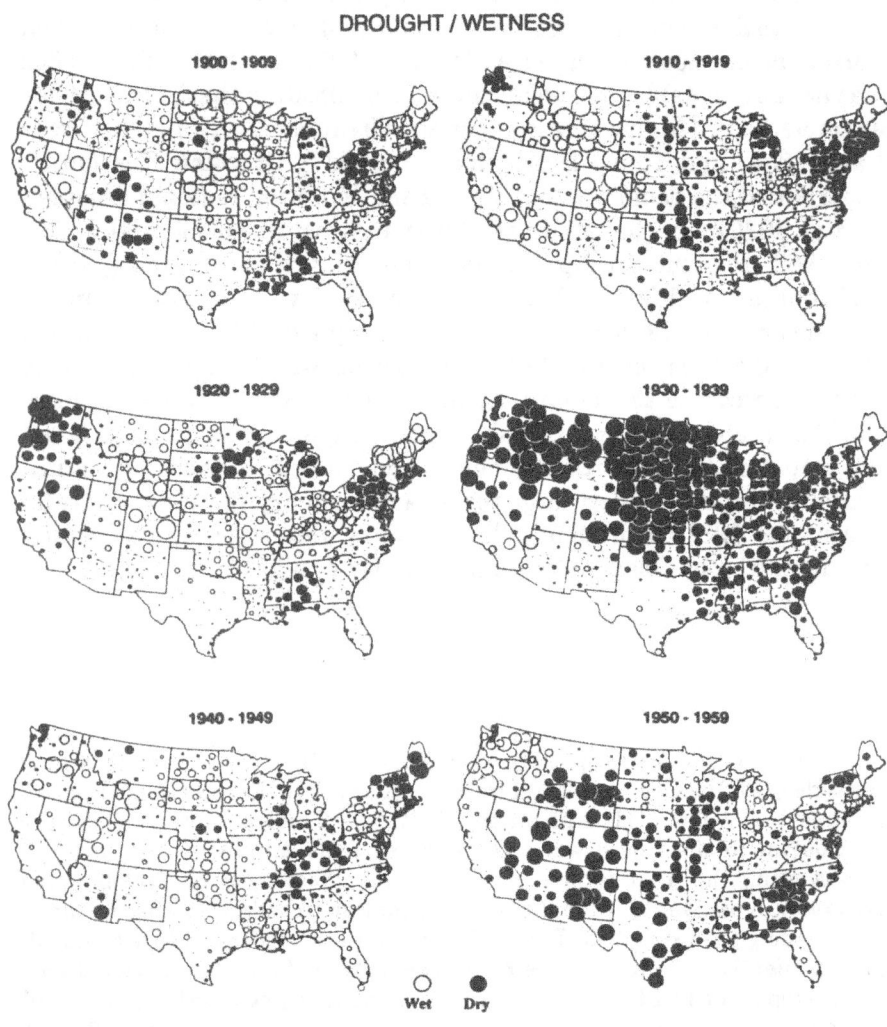

**Figure 2.25**
The severity of drought and wetness for each decade of the 20th Century as represented by the Palmer Hydrological Drought Index for each of the 344 climate divisions. Dark circles denote drought and open circles indicate wet spells, and severity of both conditions is directly proportional to the diameter of the circles (see Figure 2.3).

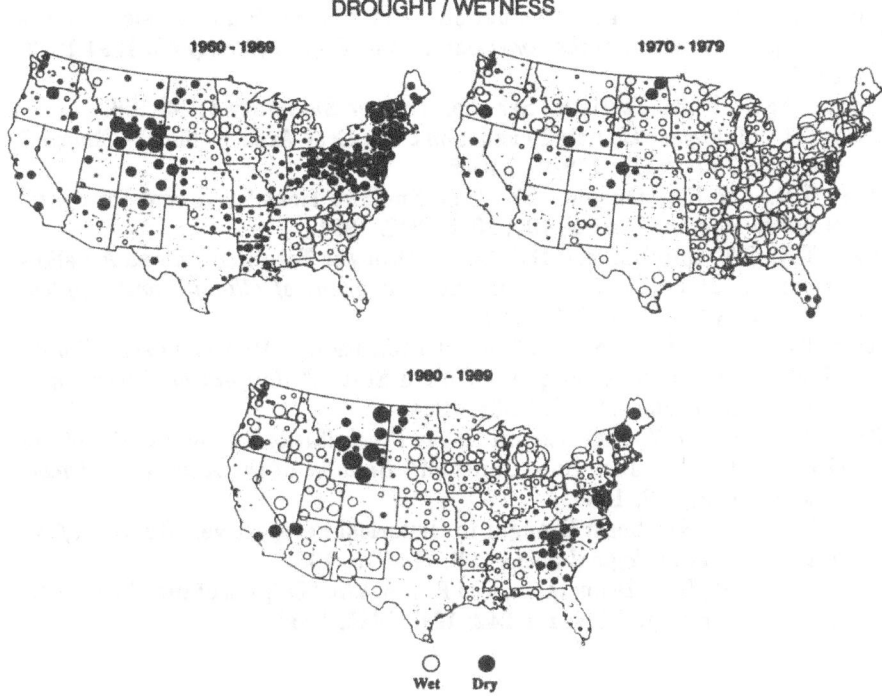

**Figure 2.25** (*continued*)

## References

Changnon, S.A. "Climate Fluctuations and Record High Levels of Lake Michigan." *Bulletin of the American Meteorological Society* 68: 1394-1042. 1987.

Illinois Agricultural Statistics Service. *Weather Summary* 4, 24. 1983.

Karl, T.R. "Multi-year Climate Fluctuations: The Gray Area of Climate Change." *Climatic Change*. 12: 179-197. 1988.

Karl, T.R., and R.G. Quayle. "The 1980 Summer Heat Wave and Drought." *Monthly Weather Review*. 109: 2055-2073. 1981.

Karl, T.R., F.T. Quinlan, and D.S. Ezell. "Drought Termination and Amelioration: Its Climatological Probability," *Journal of Climate and Applied Meteorology* 26: 1198-1209. 1987.

Karl, T.R., and W.E. Riebsame. "The Identification of 10- to 20-year Climate Fluctuations in the Contiguous United States." *Journal of Climate and Applied Meteorology*. 23: 950-966. 1984.

Karl, T.R., and P.J. Young. "Recent Heavy Precipitation in the Vicinity of the Great Salt Lake: Just How Unusual?" *Bulletin of the American Meteorological Society* 67. 1986.

———, "The 1986 Southeast Drought in Historical Perspective." *Bulletin of the American Meteorological Society* 68: 773-778. 1987.

Trenberth, K.E., G.W. Branstalon, and P.A. Arkin. "Origins of the 1988 North American Drought." *Science*. 242: 1640-1645. 1988.

# 3

---

# National Drought Vulnerability

## Section I:
## Severe Drought Conditions in 1988

The pervasive physical, social, and economic impacts of the drought, which started in 1987 and peaked in the summer of 1988, will reverberate throughout the U.S. environment and economy for some time to come. The drought pointedly reminded scientists, the public, and policy makers how sensitive environmental and socioeconomic systems are to lack of normal rainfall. Major impacts occurred in agriculture, water resources, transportation, recreation, wildlife, and other U.S. environmental and economic systems, though losses in some areas were balanced by gains in others.

The president's Interagency Drought Policy Committee (1988) estimated the total drought losses in agriculture alone during the last three-quarters of 1988 at $13 billion of direct gross national product (GNP). Second- and third-quarter GNP growth was reduced 0.9% and 0.6%, respectively, due mostly to reduced agricultural production that consequently increased retail food prices in the U.S. by 0.5%. When these figures are combined with losses in energy, water, ecosystems, and other sectors of the economy, the cost of the drought totals roughly $39 billion, making it the most expensive natural disaster ever to affect the nation.

In comparison, estimates of the worst-case hurricane that could affect the country suggest losses of roughly $7 billion (Catastrophic Losses Committee 1986) and the worst earthquake scenario thus far described would cost perhaps $30-50 billion (Lecomte 1989). Hurricanes and earthquakes are intense and relatively short-lived phenomena, and though droughts do not topple buildings, they can affect huge areas (essentially all of the contiguous U.S. was dry in the mid-1930s; in 1988 over 70% of the country experienced at least mild drought conditions) and impact essentially all aspects of economic and social life through direct or indirect effects, as described in this chapter.

## Overview

The greatest economic losses from the drought were in agriculture, where more than $15 billion in crop losses occurred. There were 20-50% reductions in corn, soybean, and spring wheat production, and the high value specialty crops of the east and west coasts were also detrimentally affected. Even though the full dimension of these impacts will never be fully assessed, we can anticipate detrimental influences on seed development, the livestock industry, and consumer food prices. Still, it could have been worse. Even though crop prices soared in 1988, major grain surpluses accumulated from prior years prevented shortages in 1988-89.

The heat wave associated with the drought of 1988 was extensive—temperatures were the highest on record over 13% of the nation, including the major metropolitan areas of the Midwest and Northeast. The heat caused an estimated 5,000 to 10,000 deaths, though the wide range of this estimate testifies to how poorly this critical effect is monitored (Avery 1988).

The environmental effects will be the most long-lasting of the drought of 1988, and include major reductions in water supplies and diminished water quality in streams and wetlands. Forest fire damage in the West was the greatest on record, and populations of certain species of wildlife in the Mississippi River Basin were reduced from 5% to 30%. There are few, if any, winners from the environmental losses.

Rapidly falling river levels in the Mississippi River Basin stopped barge traffic in June and July of 1988 and reduced barge shipments by 50% throughout the summer (see Chapter 6). This sizable obstacle created transport disruptions and price increases for the shipment of bulk commodities such as coal, grains, and petroleum products.

The drought in the central U.S. was sufficiently short-lived that effects on urban-industrial water supplies in the Midwest were mostly minor. However, it was the second year of a growing drought-induced water supply shortage in the Pacific Northwest, and the fourth year of a continuing drought in the southeastern U.S.; both areas experienced serious water supply problems.

A fifth major area to be affected was government operations. Local, state, and federal government agencies were strained by the need for added services, relief funds, and technical assistance. At the federal level, the government responded with a $4 billion drought relief bill and made $3 billion in insurance payments. Various levels of government also dealt with countless drought-related controversies, including those surrounding the management of forest fires in the West (Chapter 7) and the proposed diversion of waters from the Great Lakes to enhance Mississippi River flow (Chapter 4).

As in all droughts, there were economic "winners." Farmers growing grain and specialty crops in nondrought areas (the Deep South, southern Great Plains, and the Southwest) achieved higher profits as prices rose. Railroads in the Midwest recorded major profits from diverted river-system shipments. Well drillers were in high demand when rural and small urban water system managers realized how vulnerable they were. However, though we cannot estimate with any confidence the proportion of the $39 billion total losses that was compensated, we believe the winnings to be small relative to the losses.

## National Dimensions of Impacts: Winners and Losers

Four principal sources of information on drought impacts were used to compile this national impact assessment. First, information was gathered from the many briefings, papers, and talks presented at meetings and workshops conducted during the fall and winter of 1988-89. Second, information was gathered from newspaper reports collected during and after the drought. Third, comments were collected from interviews with persons in the public and private sectors, especially planners and decision makers, directly affected by the drought. Fourth, information has been garnered from publications, primarily those of the federal government, including reports on natural resources and economic conditions, and special drought publications issued from June onward (Changnon 1989).

We stress that the values (dollar and otherwise) presented here are estimates and are not derived from detailed economic or environmental analyses of raw data; time did not allow such in-depth analyses. One must recognize that, in many instances, these values are imprecise and subject to error of unknown size.

Table 3.1 itemizes many of the drought impacts we have identified, illustrating the wide diversity of its effects. The sectors most affected were agriculture, human health, and the environment. Agriculture was especially impacted because the intensity and areal extent of the drought was greatest during the growing season of most primary crops, May-August 1988.

Table 3.2 itemizes some of the sectors and activities that profited from the drought. The role of countervailing benefits is best illustrated in agriculture, where the U.S. Department of Agriculture (USDA)

## Table 3.1
## Roster of Drought Impacts

### A. Environmental

1. Wildlife - reduced populations, food loss for migration
2. Forests - major losses; fires, some growth stunted, seedling mortality
3. Fish - major losses in low streams, and poor quality water
4. Soil - increased wind erosion, especially northern Great Plains
5. Water - reduced quality; low, warm flows; unable to handle industrial discharges and agricultural pollution
6. Insects - some populations increased

### B. Human Health (physical and mental)

1. Deaths - number of persons totally or partly attributed to heat is in thousands
2. Illnesses - increased asthma, heat stress
3. Emotional problems - anxiety over heat stress, loss of income, higher costs for cooling, loss of recreational opportunities, concern over climate change

### C. Agriculture

1. Surpluses reduced
2. Prices up for corn, soybeans, and wheat
3. Farmers in drought areas hurt, those elsewhere helped economically
4. Long-term impacts difficult to assess due to subsidies for exports and production
5. Means to adjust to continuing drought available
6. Commercial forestry hurt
7. Increased crop insects and enhanced pesticide spraying

### D. Transportation

1. Rivers - barge traffic hurt
2. Railroads - enhanced
3. Great Lakes - shipping increased
4. Airlines - fewer weather delays

### E. Power Generation

1. Record consumption of electrical power
2. Hydropower generation reduced, costly fossil fuel substitutes required

**Table 3.1** (*Continued*)

3. Brownouts, damaged electrical equipment, discomfort
4. Increased income to most power companies

**F. Commerce and Industry**

1. Rain insurance hoax
2. All-weather peril insurance overwhelmed
3. Recreation industry received less revenue
4. Construction - fewer delays
5. Shippers - higher costs

**G. Urban Areas**

1. Reduced water supplies
2. Increased sickness and death of elderly from heat
3. Increased water consumption
4. Developed conservation procedures and penalties

**H. Water Resources**

1. Low streamflows
2. Lowered Great Lakes, reservoirs, and farm ponds
3. Lowered groundwater levels
4. New sources developed - wells drilled, piping for diversions
5. Increased costs for water and sewage treatment
6. Increased public awareness of water value and need for conservation
7. Interstate conflict heightened

**I. Education**

1. School hours reduced by heat

**J. Government Operations**

1. Establishment of drought task forces
2. Increased services and costs to government: river channeling, fire fighting, relief payments, etc.
3. Concern over $CO_2$ as cause of drought
4. Conflicts between states, especially over water
5. National attention to planning for future droughts
6. New legislation for drought relief

**Table 3.2**
**Roster of Winners from Drought Conditions**

A. **Agricultural producers** in nondrought areas and those with large surplus stocks

B. **Railroads**

C. **Water-producing technologies** (well drillers, weather modification companies, evaporation suppressant manufacturers)

D. **Electric utilities** (increased power sales)

E. **Coal companies** (increased sales from greater demand on coal-fired utilities)

F. **Great Lakes ports** (15% increase in shipping)

G. **Construction industries** (increased profits due to fewer rain stoppages)

H. **Commercial aviation** (improved performance with fewer weather delays)

---

estimated that net agricultural income in 1988 would be $57 billion—almost exactly the total income received in 1987. This estimate, despite 1988 crop losses of about $15 billion, reflects three general factors. First, producers of specialty crops, corn, soybeans, wheat, and cotton in areas that escaped the drought (portions of the South, southern Great Plains, and Southwest) had average to above-average yields. With increased prices, they experienced major income gains. Second, some farmers and most grain companies in drought areas sold surpluses accumulated in 1986 and 1987 at 1988's higher prices, helping to ameliorate their physical crop losses. Third, farmers with irrigation in drought regions also were able to sustain high yields and consequently were beneficiaries of increased commodity prices.

As mentioned before, railroads in the Midwest benefited because of the reduced shipment of bulk commodity goods on the Mississippi River system. The estimated additional income of the railroads operating in this area was $200 million.

Those dealing in "water technologies" were also beneficiaries, including well drillers, companies providing weather modification services, and companies providing chemicals for evaporation suppression.

The electrical utility sector experienced general income increases due to record high temperatures from June to August, prompting increased sales of power for air conditioning. Coal-fired utilities realized further profits as the generation of hydroelectric power was reduced by low river flows, and hydroelectric-based utilities were forced to purchase power from coal-based utilities. This led to other winners —the coal companies that had increased sales to these utilities.

Another group that benefited was the Great Lakes ports and shippers. The diversion of grain and commodities export grain shipments to the railroads led to increased movement of these grains through Great Lakes ports and a corresponding decrease in shipping from Gulf ports.

In general, all weather events of consequence produce winners and losers, but the net effect of such a drought is likely to be negative, given the physical damages and costs of alternatives.

## Impacts by Economic Sector

The effects of the drought on the nation's economy in 1988 can be disaggregated by sectors, including agriculture, transportation, power generation, recreation, and business and commerce.

### *Agriculture*

The impacts of drought, especially on agricultural systems, extend across multiple economic sectors and propagate from local to national scales (Figure 3.1). In this section we describe direct and indirect impacts of drought that expand from the farm level to the global food system.

The summer peak of the 1988 drought produced sizable impacts during the growing seasons for most crops (other than winter wheat). Extremely warm and dry conditions from May to August in the Midwest seriously affected corn and soybean crops and was widely covered in the national media. As a result, corn production dropped 45% below the 1985-87 average. Soybean production decreased 26%, and other grain crops, including barley, grain sorghum, and oats experienced 50% or greater reductions (see Table 3.3).

Wheat production was not as seriously impacted because yields of winter wheat, which produce more of the national yield than the spring wheat of the northern Great Plains, were only slightly reduced since the crop matured before the drought intensified. By comparison, spring wheat production, which was severely diminished by the drought in the central and northern Great Plains, decreased by 54%.

Table 3.3
Analysis of Effects on National Crop Production
(millions of bushels)

|  | Average 1985-87 (millions/bu) | 1988 (million/bu) | 1988 as % of Average | Reduction from 1985-87 Average |
|---|---|---|---|---|
| Corn | 8,064 | 4,462 | 55 | 45 |
| Barley | 576 | 287 | 50 | 50 |
| Grain sorghum | 933 | 540 | 58 | 42 |
| Oats | 427 | 206 | 48 | 52 |
| Soybeans | 1,981 | 1,472 | 74 | 26 |
| Wheat | 2,207 | 1,810 | 91 | 9[a] |

[a]Spring wheat was -54%

The 1988 drought caused a 31% reduction in the production of all U.S. grains. The yields of other crops were also reduced, and included a 10% decrease in fruits and vegetables. However, the sizable decreases in U.S. grain harvests were offset by large surpluses from previous years. By the end of 1989, U.S. surpluses of grains had been reduced by 60%. Sales of these surpluses moderated the effect of the drought on prices.

The effects of diminished yields and decreased production went far beyond reduced national surpluses. Commodity markets responded with increased prices of all grains, cotton, and hay. Initially, prices of corn increased 50%, from a general average of about $1.90 per bushel in early 1988 to $2.85 by the end of June 1988. Corn prices reduced after peaking in June, but remained about 35% higher throughout 1988. Additionally, wheat prices increased 30-40%, cotton prices rose 15%, soybean prices climbed 45%, and pasture hay prices went up 25%.

The livestock industry also faced the consequences of drought —prices of cattle were reduced as more animals were sent to market. Sales were up 5%, as livestock owners culled their beef and dairy cattle herds because of higher feed/grain prices, insufficient feed, and poor grazing conditions. In July 1988, the U.S. experienced the worst pasture-hay conditions ever recorded.

Farm income experienced considerable redistribution across the United States. National figures are misleading in terms of actual

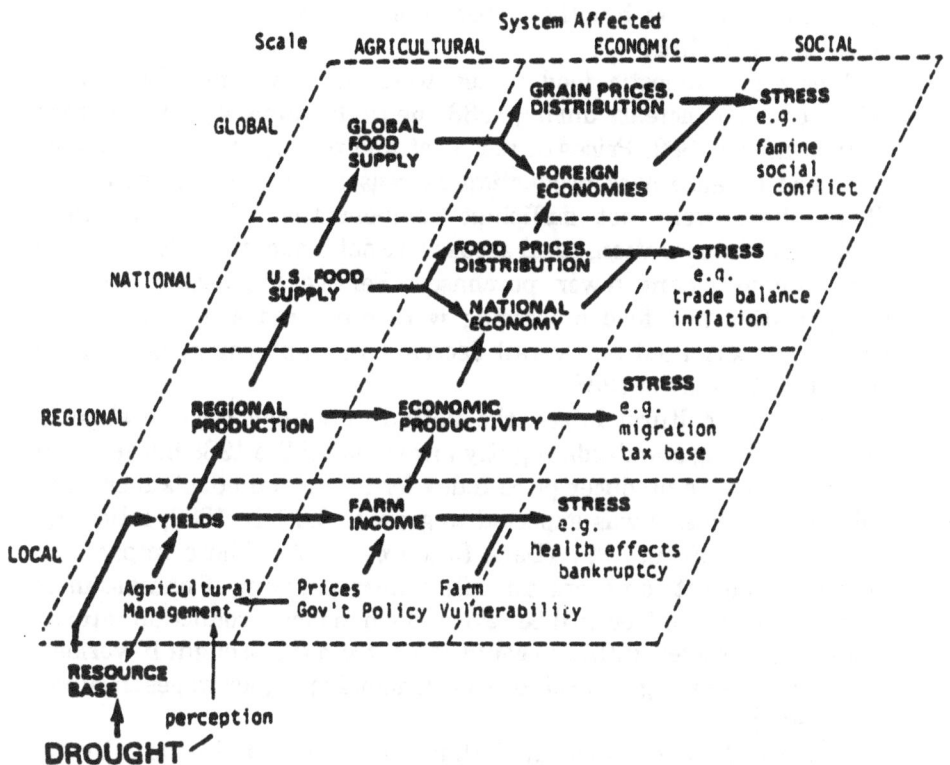

**Figure 3.1**
Impacts of drought losses across multiple systems (*Source*: Warrick and Bowden 1981)

drought losses experienced in many regions. In Illinois, for example, the average loss per farm was $48,000. The amount of reduction in farm income in all drought areas depended upon the direct loss, the amount of insurance coverage (only 20% of U.S. farms had weather insurance in 1988), and, subsequently, on farm relief payments. In response, livestock producers tended to sell more animals, and farmers sold off stored crops at higher prices to gain income. Because farmers and livestock producers were able to compensate for their drought losses with these sell-offs, gross cash receipts for 1988 were $57.4 billion, slightly larger than those in 1987.

The international trade markets saw a decline in exports from the U.S., dropping from 98 million tons of grain in 1987-88 to 87 million tons in 1988-89. A corresponding rise in unit export costs, coupled with increased production by other nations of corn, soybeans, and wheat, added to the decline.

Effects on domestic food prices were also evident. Economists estimated a 1% increase during 1988 due to the drought, with another 2% increase in 1989. Prices of red meat and poultry were up 3% over 1987, and because pork production decreased, pork prices increased. Where there were not sufficient compensating efforts or direct government intervention, rural areas were detrimentally affected. Less farm income meant fewer purchases. Service businesses such as transportation and food processing were hurt, and some rural communities experienced an overall decrease in employment as high as 20% (Data Resources 1988).

Due to agricultural losses, the nation's economy became slightly poorer and living standards slightly lower than if the 1988 harvest had been normal. The producer price index for all farm products increased to 6.2% in 1988 and was expected to grow to 9.9% in 1989, following a decline of 2.8% in 1987 (Data Resources 1988). These impacts in themselves do not represent an inflationary problem, if one assumes the relative price of food will be reversed when farm production returns to normal. However, inflation could result from the drought if workers attempt to pass on higher food costs by demanding higher wages in subsequent years.

In short, 1988 demonstrated that the nation, and especially its agribusiness sector, is still sensitive to pronounced deviations in production caused by water scarcity, despite decades of rapid advance in agricultural technology. Relatively large inventories, particularly of corn, averted major domestic food shortages in 1988, but short supplies of soybeans created significant problems within the soybean industry. Production levels in competitor nations, and, of course, U.S. weather conditions, are key to determining if more critical food problems emerge in subsequent years.

Agribusiness suppliers were also affected in a variety of ways by the reduced yields, inventory, and farm incomes and by rising food prices. Because the extent of drought was not known prior to planting, the amounts of chemical, seed, and fertilizer applied in 1988 were not changed by the drought. Given that the acreage planted in 1989 was expected to be sharply higher than 1988, additional revenues should accrue to chemical, seed, and fertilizer producers.

For seed producers, however, additional revenues were and continue to be, to some extent, offset by higher costs of production. The drought reduced seed harvests, causing an increase in the price of seeds. This had a marked impact on international markets and led to higher costs of production in tropical and southern hemisphere areas during their 1988-89 growing season.

In the U.S., sales of farm equipment continued to suffer from the financial problems that befell American agriculture throughout much of the 1980s. Some equipment sales strength had been noted in early 1988, but as the seriousness of the drought was realized in midsummer, implement sales slumped sharply.

Reduced yields of grain and soybeans led to reduced revenues for grain elevator operators and merchandisers of these commodities. These reductions were particularly significant within the soybean complex. Consequently, the process by which soybean supplies are rationed will greatly determine the ultimate impact on U.S. soybean processors.

Effects on the livestock processing sector were buffered by the relatively small size of cattle herds maintained by producers. Herd liquidation was not a major phenomenon, and with adequate grain reserves, beef processing firms were not greatly affected by the drought.

*Transportation*

Inland waterway transport experienced severe repercussions from the 1988 drought (see Chapter 4). The transport of grain contributes about 10% of total U.S. rail tonnage and about 10% of inland waterway activity. Loadings destined for export markets account for nearly 40% of total grain movement by railroads and a majority of all barge movements, and because of sensitivity to weather and economic conditions, grain transport is often a major factor in the health of these two industries.

Roughly 300 barge and tow companies operated on the Mississippi River system in 1988. Collectively, these companies carried 40% of all grain products transported in the United States. Beginning in June, record-low flows on the lower Mississippi River (south of Cairo, Illinois) limited normal barge passage. Intensive dredging was conducted by the U.S. Army Corps of Engineers to open 11 areas where blockages occurred. As the problems grew, barge shipping prices doubled, then tripled, making railroad shipping charges competitive with barge prices in the central United States. Shipment of bulk commodities (coal, petroleum, and grain) on the Mississippi River system dropped by 50% throughout the summer of 1988, resulting in losses of over $200 million, or 20% of the annual income of the river transport industry (Changnon 1989).

Some railroads anticipated the problem, leased additional railcars to handle the increase in grain and coal shipments, and realized sizable profits during the summer and fall. Because of this shift in transportation, Great Lakes ports experienced a sizeable increase in grain exports, while the ports on the Gulf of Mexico suffered serious losses. Total losses to the barge and tow industry, coupled with increased costs to shippers, was about $1 billion.

*Power Generation*

Record high temperatures through much of the central United States, coupled with widespread above-average temperatures elsewhere in the nation, led to marked increases in the use of air conditioning from late June through August of 1988. Electric power demand reached record levels in the central and northeastern U.S., increasing income to power utilities. Delays in barge movements on the Mississippi River increased the cost of shipping coal, leading to coal price increases for some utilities.

The generation of hydroelectric power on the Missouri River, in the Pacific Northwest, on the Ohio River, and in the Southeast (including Tennessee Valley Authority and Corps of Engineers reservoirs) was reduced between 20% and 40%. Overall, national hydroelectric power generation in 1988 was 13% below 1987 levels. Companies dependent upon hydroelectric power had to purchase power from coal-fired utilities at higher prices, increasing their operating costs by $4.2 million. Two cases in particular underscore the difficulties: (1) the Power Marketing Administrations of the Department of Energy had to purchase power from others, and (2) by July 15, the Tennessee Valley Authority was able to produce only 53% of its normal hydroelectric output and therefore experienced a loss of $150 million (Interagency Drought Policy Committee 1988).

Further problems were caused by low river flows and lack of cooling water in many midwestern power plants. High water temperatures also reduced the efficiency of cooling, causing generating plants to close along the Mississippi and Tennessee rivers. Salt intrusions into the Mississippi River Delta, caused by reduced flows, affected major petroleum refineries at New Orleans—fresh water had to be barged in to operate boilers and cool machinery.

*Recreation*

Outdoor recreation, a sizable and growing economic sector, was hurt by the drought in several ways. First, lowered lake and river levels negatively affected water recreation facilities and reduced pleasure boating in many areas. Second, low flows and poorer water quality harmed waterfowl and fish, leading to fish kills and reduced amenity values, especially in the central U.S. These difficult environmental conditions reduced the populations of other forms of wildlife, which in turn affected fall hunting seasons. As a result, for example, reductions were made in the length of hunting seasons and bag limits for deer and other wildlife normally hunted in midwestern, western, and southern states (Interagency Drought Policy Committee 1988).

Extreme fire danger in the West brought outdoor recreation almost to a halt in some areas. Montana's governor legally banned all outdoor recreation in August and September of 1988. TW Services, the tourist facilities concessionaire in Yellowstone National Park, lost several million dollars in revenues (no actual amount is available), and similar losses occurred throughout the West.

*Commerce and Industry*

Impacts on American commerce and industry were mixed. Businesses manufacturing and selling air conditioners and swimming pools profited, while many recreation businesses suffered from a decrease in the number of vacations taken by Americans because of hot weather, fire danger, and reduced personal income. The agrichemical industry was expected to see a stronger demand for fertilizers and pesticides due to greater acreage planted. The lumber industry suffered timber losses from fires and declines in future productivity from seedling losses. Conversely, the construction industry benefited by the large number of work (nonrain) days, leading to completion of projects on or before schedule. Similarly, the airline industry experienced a record low number of flight delays because of the lack of wet and stormy conditions during 1988.

The major business losses and costs of the 1988 drought are listed in Table 3.4. Most losses and costs were related to lost agricultural production, valued at $15 billion, and payments to farmers of $7 billion for relief and insurance. The banking industry also suffered losses in areas where farm income was reduced and there was no concurrent government aid, but remained sufficiently strong in areas that received aid to prevent debt service payments from being threatened nationally. Banks in the hardest hit areas, including Minnesota, the Dakotas, Illinois, and Wisconsin, experienced increased debt service problems (Data Resources 1988).

The U.S. industrial sector experienced limited effects from the 1988 drought. At the macroeconomic level, tighter food supplies generated higher prices and subsequently higher interest rates, which slowed

## Table 3.4
## Losses and Costs of the 1988 Drought

| Expenditure/Loss/Cost | Amount |
|---|---|
| Federal Disaster Assistance | $4.0 billion |
| Federal Crop Insurance (plus emergency feed assistance) | 3.0 billion |
| Transportation | 1.0 billion |
| Agricultural Output (overall farm production) | 15.0 billion |
| Energy Production Costs (hydropower and coal) | 0.2 billion |
| Food Costs | 10.0 billion |
| Forests | 5.0 billion |
| Agricultural Services | 1.0 billion |
| Total | $39.2 billion |

economic growth slightly and shifted it away from consumer goods. Producers of processed foods and beverages suffered. The higher 1988 grain prices had little effect on food consumption, because they were offset by lower meat prices in late 1988 and early 1989. However, an 8% jump in prices of both meat and nonmeat processed foods was predicted for 1989 (Data Resources 1988). One expected reaction was that Americans would eat more often at home and trade down from more processed (and more expensive) foods to more basic foods. Economists estimated that processed food output would rise 2.7% in 1988, but only 1.1% in 1989—substantially less than the 3.4% average growth that had been projected before the 1988 drought. Such a slowed growth in processed food production would reduce packaging requirements and lead to reduced manufacturing of paper containers, cans, and plastics.

The service-related sectors of regional and state economies bore significant impacts from the 1988 drought. States in the upper Midwest experienced substantial reduction in retail trade and service jobs in areas where gross farm income fell. Drought-induced weaknesses in services employment in 1989 were chiefly limited to those states that substantially reduced their livestock herds in 1988-89. Conversely, there was added employment for fire fighting in the West. Towns around Yellowstone National Park recovered some of their lost tourist income by housing and feeding fire fighters.

The drought's impacts on transportation and wholesale trade employment were subtle and complex. As stated earlier, the drought altered the transportation mode and destination of agricultural goods. Railroads benefited and barge operators suffered, but both were hurt by the slowdown in U.S. grain exports. Impacts in the wholesale industry can only be guessed at—increased demand for some materials and manufactured goods, such as air conditioners, and reduced demand for others.

The drought also slowed retail sales. By reducing harvests and raising food prices, it was expected that the drought would depress real income growth from 2.5% in 1988 to 1.7% in 1989. As a result, the growth of real consumption nationwide was expected to slow from 1.9% in 1988 to 1.4% in 1989 (Data Resources 1988).

The drought also widened regional disparities in income growth and business opportunities. The central regions, hardest hit by drought, were the same regions that experienced the slowest growth through the 1980s. Retail sales experienced the greatest declines in the northern Great Plains and western Corn Belt, where sales volume fell 2.5% in 1988 and was predicted to fall 1.6% in 1989.

The drought also slowed the promising pre-1988 recovery in U.S. agricultural exports (Data Resources 1988). The benefits of higher food prices from the drought were overshadowed by supply-driven declines in export volume, resulting in a shrinking agricultural trade surplus in 1989. This setback occurred just as the U.S. started to regain export markets lost earlier in the decade. Agricultural producers in Europe, South America, and Australia capitalized on the U.S. drought by boosting their exports to meet the world's supply shortfall, placing the American farmer at a competitive disadvantage.

It was estimated by the Federal Drought Committee that the approximate total cost of the drought was $39 billion. Yet, even this large loss, which may be an underestimate considering that many impacts are not readily counted in dollars, resulted in a gross national product downturn of only 0.4% below expectations. The Consumer Price Index went up 5% in 1988 and government economists attributed 0.3% of that climb to the drought. Thus, in terms of the national economy, the drought's economic impacts were not monumental. Yet, they are undoubtedly the nation's largest aggregate impacts due to adverse climate conditions in the 20th Century, and the apparent diminishment of total losses when compared to the national economy neglects the plight of smaller economic units—states, companies, farms, or individual households—that experienced sizable losses.

## Personal Income and Cost of Living

Agriculturalists in states unaffected by the drought, such as California and Florida, actually benefited from the drought by receiving higher prices for their products. By contrast, those living in the western Corn Belt, northern Great Plains, and Northwest were dealt sharp blows to personal income from the loss of crop production.

Consumer food price inflation was modest in 1988 because the farm-level prices of most crops account for only a small share of the retail value of finished products. There are a few products, like pasta, that became more expensive because they were especially sensitive to spring wheat losses. However, since farm-level prices constitute nearly half the retail value of meat products, the impact of the drought on meat prices was more sizable. In the near term (late 1988-89), meat prices weakened because of distress sales by farmers. However, livestock producers had to contend with reduced quality rangelands and increased prices of corn and soybean meal in 1989, causing substantial cost increases (see Chapter 3, Section II).

Experts expected that the drought, primarily through increased food prices, would cause a 0.1% increase in the 1988 inflation rate and 0.2-0.3% increase in the 1989 rate (Data Resources 1988). However, this forecast assumed that the drought's inflationary impact was essentially confined to the food sector. Such economic analyses do not account for altered prices of other bulk commodities, nor the costs of services and utilities.

## Selected Environmental and Ecosystem Effects

The environment and ecosystems experience the most subtle and enduring impacts of drought. Cumulative stress on wetlands, wildlife, forests, groundwater, and soils cannot be measured accurately, nor even in many cases estimated with any credibility. Indeed, many environmental effects are neglected in impact accountings, suggesting that the overall effects of drought are underestimated. Thus, we attempt at least to roster the environmental and ecosystem impacts of the 1988 drought.

### Water Resources

The impacts of the 1988 drought on water resources depended upon several factors, most notably pre-existing conditions. Although the drought was especially severe in the central U.S. during the summer of 1988, it did not create severe water supply problems there because ground water and reservoir supplies were adequate. However, Midwest communities dependent upon surface water experienced growing shortages as 1988 ended, and local conservation had begun in some areas as dry conditions continued into 1989.

Moreover, 1988 was the fourth year of drought in the southeastern United States and the second year of drought in the Pacific Northwest and California, making difficult conditions worse. In the Southeast, problems emerged in dozens of local water systems and in two federally developed multipurpose systems. Tennessee Valley Authority reservoirs entered the year below normal, and only careful water management to balance hydropower, transportation, and water supply provided adequate water for most purposes during the summer. However, operation problems in multipurpose reservoirs in the Savannah River Basin and along the Apalachicola-Chattahoochee-Flint River system, which actually began in 1985, were worsened to the point that water supplies for some uses were eliminated. In fact, averting water shortages in the Atlanta metropolitan area required near-total curtailment of hydropower generation and barge transportation. Claims of federal mismanagement of these reservoir systems were common, but there is good evidence to suggest that managers had learned and applied lessons from previous droughts and made the necessary hard choices (see Chapter 6).

*Forests*

The drought's damages to forests, both those producing lumber and those preserved in park lands or dedicated to habitat and watershed protection, are included under the environmental, rather than economic, impacts because we feel it is important to recognize the more subtle effects of natural hazards on ecosystems. The record extent of forest fires and weather stresses on young trees will surely have great economic impacts on lumber and paper industries in years to come, but the impacts on forest ecosystems may be even more profound.

Sixty-eight thousand wildfires burned 5.1 million acres of U.S. forest land during 1988, the worst since the early 1900s. This was the fourth consecutive year of severe fires, particularly in the Pacific Northwest—a period that helped prepare the nation's fire-fighting system for the 1988 onslaught. The response involved 30,000 fire fighters during July and August, required fire fighters from Canada and the U.S. military in several western states, and cost an estimated $300 million.

Outside the fire areas, extremely warm and dry conditions greatly reduced the number of tree seedlings—mortality may be as high as 40% of the trees planted nationally during the last 10 years, including 150 million pine seedlings. Drought-related stress to trees also increased insect attacks, and 5.7 billion board feet of lumber were lost to pine bark beetles. In the aggregate, the timber harvest of 1988 was curtailed by 20%, chiefly because of fire and fire dangers. Furthermore, the fire losses in 1988, coupled with the damage to young trees, are expected to negatively affect forestry for up to 20 years, regardless of whether another drought occurs.

## Fish and Wildlife

A variety of impacts occurred to the nation's wildlife, though few have been recorded in any formal fashion. Fish, waterfowl, and wildlife mortality greatly increased in the TVA reservoir system, and salmon spawning suffered in West Coast rivers. Game bird populations throughout the central and western Great Plains were hurt from a loss of grass seed caused by poor growth and wetland desiccation.

Both wetland and big game wildlife in the mountainous Northwest experienced reductions in breeding habitats. Western streams and rivers, especially in the northern Rocky Mountains, endured a worsened trout habitat, and the effects of runoff from burned areas remains to be evaluated. High water temperatures due to the warm, dry conditions in the East resulted in an increase in oyster diseases in Chesapeake Bay. The 1988 oyster harvest was 375 million bushels, the lowest on record; 975 million bushels were harvested in 1987. The reduction of nutrient runoff in the Mississippi Delta reduced the productivity of sea trout and crabs, and several endangered species were threatened by habitat reduction.

In many instances, wildlife losses are an economic, as well as environmental, impact, and more formal quantitative accounting may eventually be available. In general, however, the environmental impacts of the drought have been poorly quantified. Some of the problems arise from the lack of monitoring of the effects of weather on critical environmental elements and the difficulty in measuring the effects that will be expressed over several years. The most long-lasting effects of the 1988 drought will occur in the environmental, not the economic, sector.

## Human Health

Public officials paid little attention to one of the most significant aspects of the 1988 drought: the impacts on human health. The Center for Environmental Physiology has estimated, based on studies of a heatwave that occurred in 1980, that the number of deaths directly or indirectly related to the high temperatures of 1988 were between 5,000 and 10,000. Final federal figures are not yet available, and many deaths totally or partially caused by heat stress are not recorded as such on death certificates. Many of the deaths occurred in large metropolitan areas of the central and eastern U.S., particularly in inner cities where very young, old, or malnourished persons died in low-cost housing that lacked air conditioning and proper ventilation. Local and state agencies attempted to help with medical assistance, workshops, and warnings about the danger of heat stress, but the federal government paid little attention to this area of concern.

Another difficult to measure effect of the drought was on mental health. Mental stress due to failing crops; persistent hot, dry weather; and difficult economic times led to heightened anxiety among farmers. Anxiety clinics and workshops were organized in the Midwest during July and August for farmers and their families. Others found themselves sequestered indoors with air conditioning for much of the summer or altered their vacation and recreation patterns because of continuously high temperatures, fires in the West, and other environmental impacts.

## Public Sector Issues

Local governments were affected in several ways. Those experiencing water shortages raised water and other public utility rates and enforced conservation procedures. Municipal governments also responded to health problems by providing "cooling centers" and increased emergency medical response. Many midwestern and southern cities had developed plans for such responses during the 1980 heat wave.

Local and state governments also used various "technological fixes" to obtain more water. Many communities in the Midwest drilled new wells to tap deeper groundwater. Several states, like Ohio and California, launched weather modification projects to enhance precipitation. Some communities and local industries spread chemicals on small lakes and ponds to suppress evaporation.

Some state governments set up their own farm relief and assistance programs. Illinois provided $65 million in state aid to farmers. State agencies also assisted communities with water supply problems by providing advice and financial support for new facilities, including pipelines, to reduce local water problems. Most states in the central U.S. established drought task forces comprised of representatives of relevant state agencies (agriculture, water, environmental protection, etc.). A few had pre-existing drought plans, and in many respects, states reacted to the drought more quickly and with better planning than did the federal government, perhaps because state governments are in closer contact with their resource problems. Yet, ultimately, the

1988 drought, particularly as it related to agriculture, was so great in areal extent and intensity that it required a national, not just a state-level, response. However, there was little evidence of effective coordination between state and federal agencies.

Overall, however, the drought had only a mild impact on state and local government budgets. Increased inflation from higher food prices from the drought was expected to boost sales and income tax collections in 1989. Conversely, states in the northern Great Plains and western Corn Belt were expected to experience reduced revenue growth in 1989 due to the drought (Data Resources 1988).

The most notable federal legislative response to the drought occurred during the 100th Congress. Legislation originating in the Senate and the House of Representatives in July led to the Agricultural Relief Act of 1988, signed into law on August 11, 1988 (National Climate Program Office 1988).

## Summary

Overall, analysts saw the drought of 1988 as having a relatively minor effect on the national economy. The Interagency Drought Committee's December report to the president concluded that there will be "little effect on the overall growth rate of the U.S. economy from the drought of 1988." The collective national impacts do not seem significant when examined within the context of GNP and the Consumer Price Index, but the economic, environmental, and health impacts make the drought of 1988 one of the nation's worst natural disasters. It can be argued that several ameliorating factors, like surplus grain stores and federal agricultural subsidies, which could be changed to drought relief without impacting the federal budget, may not be as available during the next drought.

The major impacts of the drought of 1988 are summarized in Table 3.5, and the long-term and cumulative impacts are delineated in Table 3.6. Although the environmental damages are less well-known than others, they may be the most sizable and long-lasting. Economically, the agricultural sector was hardest hit—the nation lost 31% of its normal grain production, which could have precipitated a food supply crisis had there not been large surpluses from prior years. Human health impacts included an estimated 5,000 or more deaths.

The total economic losses and costs of the drought were roughly $39 billion, though we suspect that this is an underestimate, and that accounting of enduring and cumulative effects would increase the total.

**Table 3.5**
**Major Impacts of Drought of 1988**

A. **Environmental damage** - 5.1 million acres of forest burned; fish and wildlife habitat reduced, water resources and quality reduced, and waterfowl/wildlife mortality increased

B. **Agricultural losses** - 31% reduction in U.S. grain production; farmers, agribusinesses, rural communities seriously impacted

C. **Human health** - 5,000 to 10,000 deaths, anxiety among farm families

D. **Water shortages** - urban-industrial, transportation, power generation, and recreation

E. **Transport reductions** - on inland waterways, changed transportation patterns

F. **Increased cost of living** - food and electricity

G. **Electric utilities** - increased costs and profits

H. **Government** - added services, increased costs, changed policies

## Table 3.6
## Long-Term and Cumulative Impacts

A. **Forests** - major damages lasting 10-20 or more years; possible increased pest damage

B. **Environmental losses** - habitat and wildlife will recover slowly

C. **Effects on agriculture production in 1989**
 1. Production
     a. Expanded acreage
     b. Yields/production changed
     c. Carryover effects of herbicides and fertilizers applied in 1988
     d. Financial effects, altered purchasing power of farmers
     e. Insurance purchases increased
     f. Reduced carry-over stocks
     g. Change in market competitiveness
 2. Agribusinesses
     a. Income increases: fertilizer, seed, chemical firms
     b. Income losses: equipment sales, insurance, and storage (elevators)

D. **Changes in government policy**
 1. Management of farm programs
 2. Ag-export policy
 3. Forest management

E. **Shifting strategies in private sector**
 1. Railroads vs. barge transportation
 2. Utilities expansion, especially peak-load
 3. Altered prices of shipping and food products

F. **General effects**
 1. Wider demand and application of climate data and seasonal forecasts
 2. Greater interest in weather modification
 3. Greater focus on interstate water management

# References

Avery, M. Heat Death Toll in 1988. Washington, D.C.: Center for Environmental Physiology. Personal communication. 1988

Catastrophic Losses Committee. *Catastrophic Losses: How the Insurance Industry Would Handle Two $7 Billion Hurricanes.* Oak Brook, Illinois: All-Industry Research Advisory Council. 1986.

Changnon, S.A., ed. *The 1988 Drought.* CCR No. 14. Mahomet, Illinois: Changnon Climatologist, Inc. 1989.

——. "The 1988 Drought, Barges, and Diversion." *Bulletin of the American Meteorological Society* 70: 1092-1104. 1989.

Data Resources. *The Business Impact of the 1988 Drought.* DRI Special Report. Lexington, Massachusetts: Data Resources, Inc. 1988.

Interagency Drought Policy Committee. *The Drought of 1988.* Final Report of the President's Interagency Drought Policy Committee. Washington, D.C.: White House. 1988.

Lecomte, Eugene. "Earthquakes and the Insurance Industry." *The Natural Hazards Observer* 14, 2: 1. Boulder: Natural Hazards Research and Applications Information Center. 1989

National Climate Program Office. *The Drought of 1988 and Beyond.* Rockville, Maryland. 1989.

Warrick, R.A., and M.J. Bowden. "Changing Impacts of Drought in the Great Plains." In *The Great Plains: Perspectives and Prospects*, edited by M. Lawson and M. Baker. Lincoln: University of Nebraska Press. 1981.

# Section II:
# Continuing Impacts in 1989

Severe spring and summer dryness in the central U.S. brought national attention to the drought of 1988. In other areas, such as California and the Pacific Northwest, 1988 was the second year of drought. In the southeastern United States, it was the fifth year in a recurrent drought pattern that had begun in 1981. Parts of the Deep South and the Southwest that had escaped drought in 1988 and prior years (Figure 3.2a), however, were experiencing drought conditions, while some areas seriously affected in 1988 found relief in welcome rains.

The major agricultural impacts and resulting adjustments associated with continuation of the drought in 1989 fall into two categories: (1) those resulting from the drought conditions of 1988 in areas where the drought ended during the winter-spring of 1988-89, and (2) those in areas that had either continued drought into 1989 or began to experience drought during the winter, spring, and summer of 1989.

Continuation of the drought in 1989 evoked a national drought policy crisis. It was founded on uncertainty over the potential for, and nature of, continuing drought and included the climatological myth that back-to-back droughts were impossible. It was also fueled by uncertainty over how to respond, uncertainty over the interactions of various responses, and the hiatus in active federal monitoring and assessment after the fall of 1988. The drought's continuation provided an opportunity for us to extend our analysis to the effects of more persistent drought and to assess how the nation responded to a multiyear event.

## Drought Conditions in 1989

After the 1988 crops had been harvested, national interest in the drought diminished rapidly. President Reagan's task force on drought summarized the drought effects and published their final report in December 1988 (Interagency Drought Policy Committee 1988). Little top-level federal attention was directed toward maintaining drought monitoring or planning activities during the winter of 1988-89. The belief was that the problem had or would disappear and that if it returned, ad hoc management could be effectively employed again.

One basis for this optimism was the myth that back-to-back droughts were unprecedented and unlikely. While two concurrent extreme drought years are unlikely, it is well-known that U.S. droughts have occurred in multiyear episodes (e.g., the 1930s and

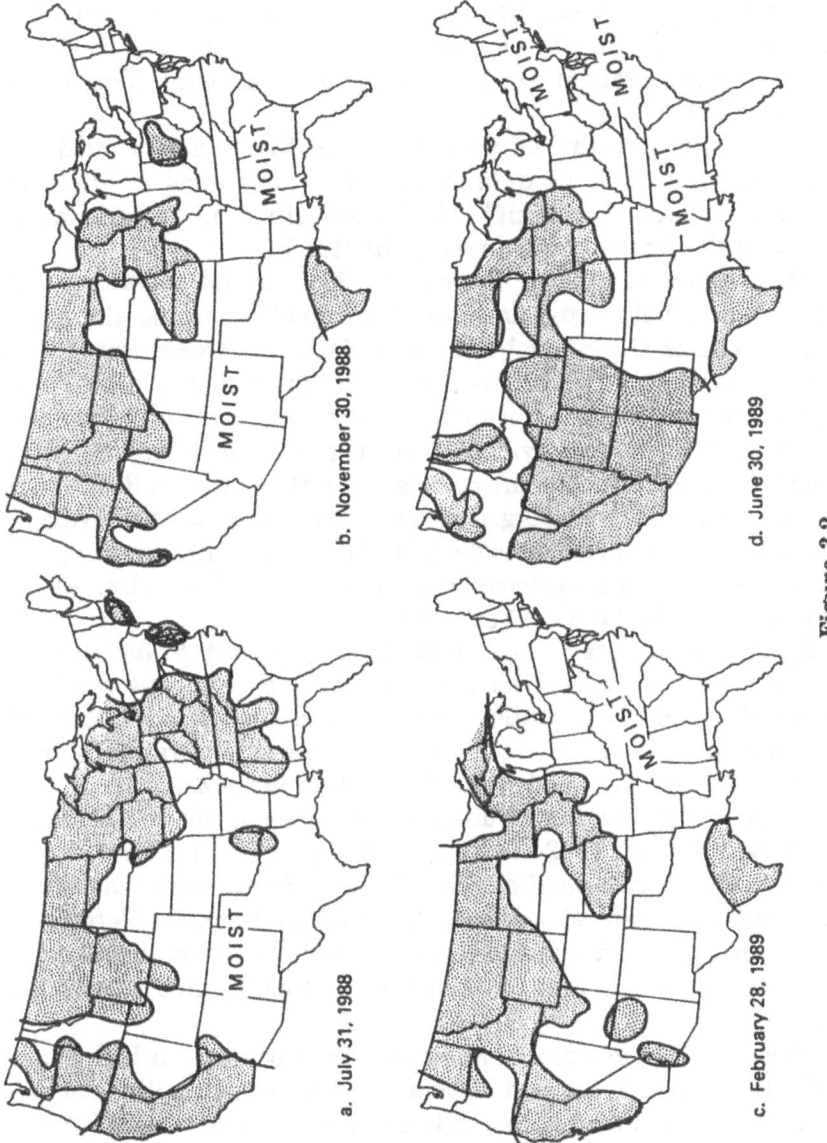

a. July 31, 1988

b. November 30, 1988

c. February 28, 1989

d. June 30, 1989

**Figure 3.2**
**Areas with severe to extreme drought**

1950s) and that, indeed, there are historical and theoretical reasons to *expect* drought to persist from one year to the next due to the "memory" of large-scale atmospheric patterns and persistence of conditions like low soil-moisture levels. Yet, the USDA stated that it did not expect serious drought in 1989 (*Champaign-Urbana News-Gazette* February 20).

In fact, the drought did persist through the winter of 1988-89 in some areas, as well as develop in many new areas in the West. Nevertheless, the first official 1989 drought situation analysis was not issued by the federal government until March.

The winter months of December 1988 through February 1989 brought, as usual, mixed amounts of precipitation across the nation. Many areas of the central U.S. and Northeast experienced below-normal snowfalls, and below-normal precipitation prevailed in the High Plains and western Corn Belt. By the end of February, four areas of serious drought were evident: (1) central California (again), (2) the central High Plains (especially Kansas), (3) the northern Rockies, and (4) the western Midwest. Agricultural fears grew over the developing soil-moisture shortage, particularly in the winter wheat growing areas of Kansas, Nebraska, western Oklahoma, and west Texas. Dryness was also emerging in the New York area.

Spring (March-May) conditions in the United States featured extremely heavy and persistent precipitation from eastern Texas through the Ohio River Valley into New York and New England. The wet, late spring, characterized by above-normal May rainfalls in the Northeast, effectively ended the feared *hydrologic drought*, or water shortage. It also terminated drought conditions in the Southeast that had persisted since 1982. The spring precipitation, however, did not alleviate the *agricultural drought* (deficient soil moisture) of the western Corn Belt or the central-southern High Plains (Nebraska-West Texas). Winter and spring precipitation levels were also deficient in central and southern California, setting those areas up for summer drought conditions (Figure 3.2d).

May 1989 in the western Midwest and southwestern U.S. ranked as the 16th driest in the past 95 years, and drought conditions in these areas at the end of May were actually more extensive than those of May 1988. Severe or extreme drought affected 31.5% of the U.S., the eighth greatest areal extent of dry conditions since 1895. Also in May, 43.8% of the Mississippi River Basin had extreme long-term drought— only the May values of 1931, 1934, 1941, and 1954 exhibited greater areas of severe drought (Figure 3.4). For the corn and soybean belts, the areal extent of extreme and severe drought in May 1989 was

41.8%, greater than 1988 and ranking as the eighth largest value for those areas.

The South and East received very heavy late-spring and June rainfalls, and areas of severe drought shifted to the west—California, New Mexico, Colorado, Utah, Wyoming, and Nevada. On a national basis, the drought conditions that began early in 1987 persisted well into the summer of 1989 (Figure 3.2d). The spring of 1989 became the fifth consecutive spring in the U.S. with below-average precipitation. The only comparable historical period of consecutive dry springs was 1913-17.

Figures 3.3, 3.4, and 3.5 demonstrate, however, that the primary locations of drought in 1988 had largely shifted to different parts of the nation by June 1989, such as the central Great Plains. The graphs for Nebraska and Kansas (Figures 3.6 and 3.7) show that the precipitation values from winter 1988 to spring 1989, the period of winter wheat growth, were *extremely* low—the lowest on record for Nebraska and the third worst for Kansas.

## 1989 Drought Impacts

Continuation of the drought into the first nine months of 1989 had great consequences for agriculture, reflecting a continuing financial instability among individual farmers, agribusinesses, and the farm markets. Agricultural difficulties led to renewed governmental attention to drought, including debate over several relief proposals and the successful implementation of a few remedies.

The effects of the continuing drought in 1989 are examined here in terms of (1) the individual level, including Great Plains wheat farmers, livestock producers, and the midwestern corn and grain farmers; (2) the grain market level; (3) agribusiness in general; and (4) governmental reactions.

### Great Plains Wheat Farmers

One of the three areas most impacted by the drought in 1989 was the winter wheat belt in Kansas, Nebraska, Oklahoma, and west Texas, which experienced conditions approximating the "worst case" for winter wheat. Problems began as the drought intensified during the dry fall of 1988, followed by an unusually warm December and January that ended abruptly with an outbreak of extreme cold in February—a combination leading to "winter kill" of the young wheat plants. Late winter and early spring dust storms further damaged the wheat. Finally, recurrent above-normal temperatures (in the 90°F range) during April

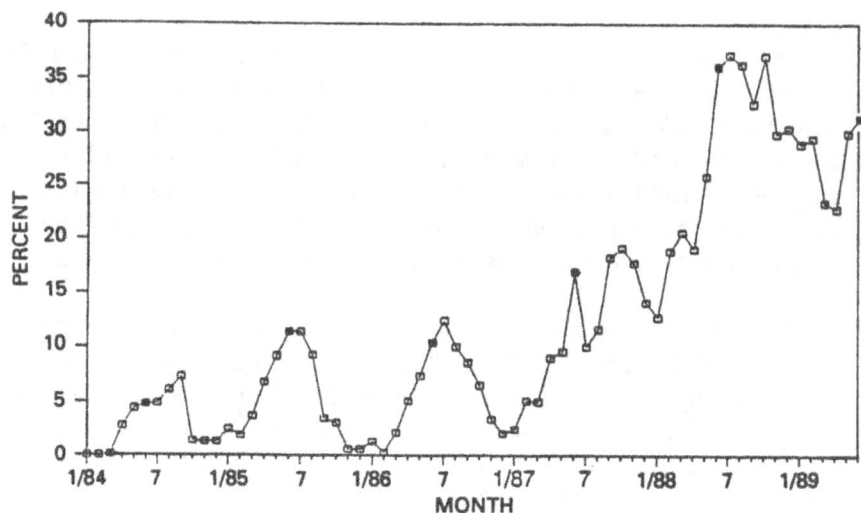

**Figure 3.3**
Percent of U.S. with severe to extreme drought (*Source*: National Climatic Data Center)

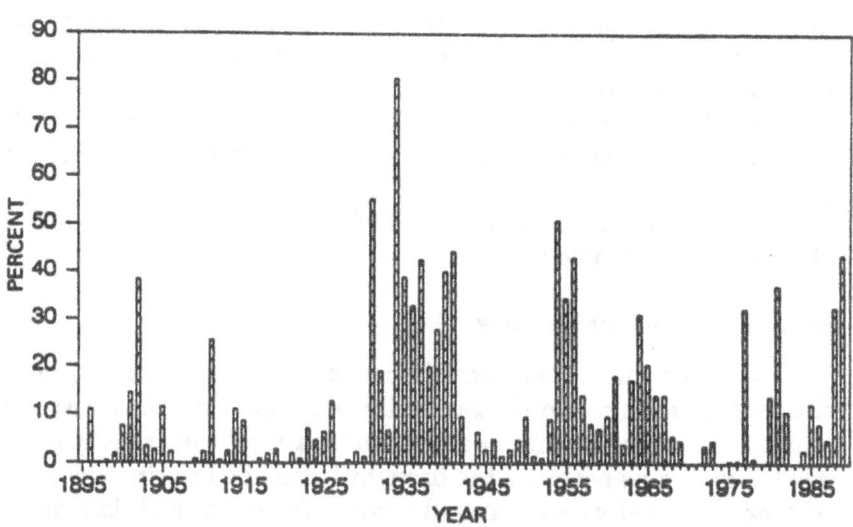

**Figure 3.4**
Percent of Mississippi River Basin with severe to extreme drought in May, 1895-1989 (*Source*: National Climatic Data Center)

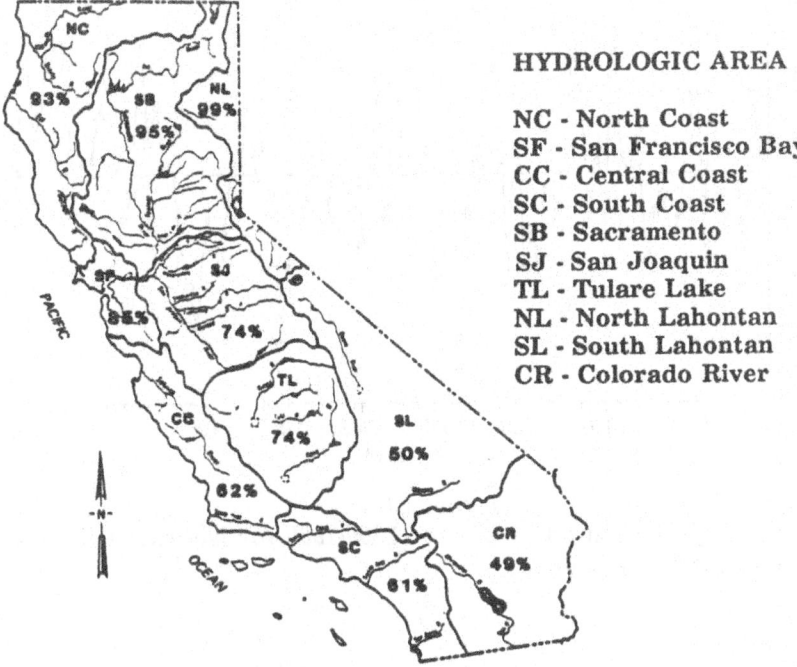

HYDROLOGIC AREA

NC - North Coast
SF - San Francisco Bay
CC - Central Coast
SC - South Coast
SB - Sacramento
SJ - San Joaquin
TL - Tulare Lake
NL - North Lahontan
SL - South Lahontan
CR - Colorado River

Figure 3.5
Percent of average precipitation in California basins for October 1988-
April 1989 (*Source*: Western Regional Climate Center)

and May, accompanied by low precipitation, further damaged the
maturing wheat crop (Figures 3.6 and 3.7).

The USDA reported in January that winter wheat planting in the
fall of 1988 was down 12% from prior years (*News-Gazette* January 17,
1989). This report echoed the apprehensions of Great Plains farmers
concerning low soil-moisture content as a result of the dry conditions
during the fall of 1988 (Figure 3.2b).

Problems with the hard red winter wheat crop made national news
by early April, when it was announced that at least 25% of the Kansas
wheat crop had been lost to the drought, with similar losses in west
Texas, Oklahoma, eastern Colorado, and Nebraska (*Los Angeles
Times* April 10). Economists speculated that bread prices would rise 10%
during 1989.

Political attention to the problems in the High Plains also emerged
in early April. Congress and the USDA began to debate the severity
of the drought and potential adjustments to its impacts (see section on

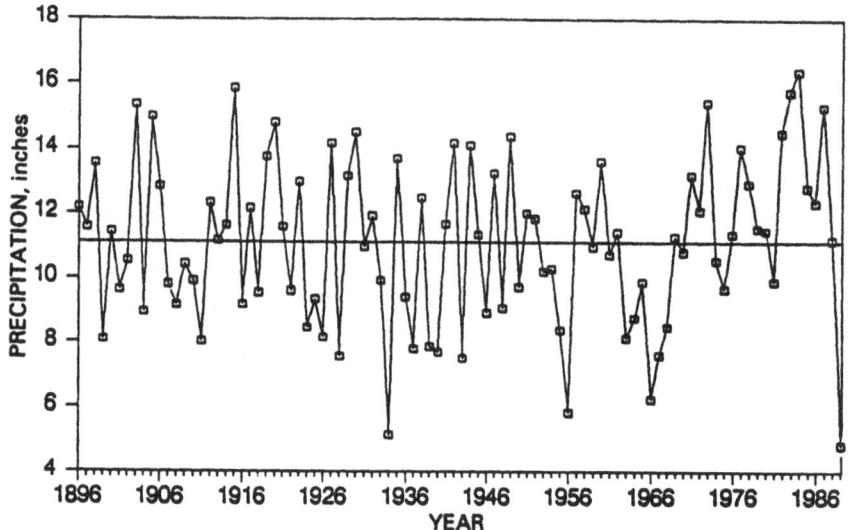

Figure 3.6
October-May precipitation values for 1895-1989 for Nebraska (*Source*:
National Climatic Data Center)

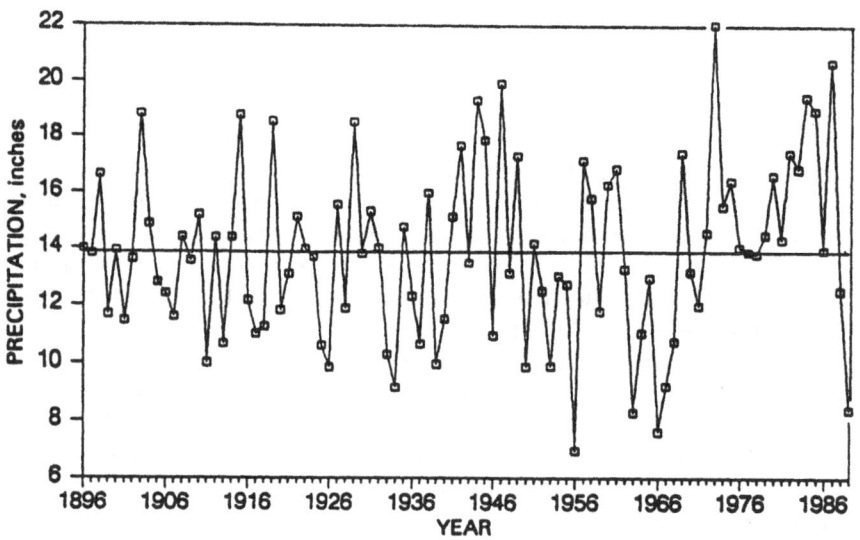

Figure 3.7
October-May precipitation values for 1895-1989 for Kansas (*Source*: National
Climatic Data Center)

*Governmental Reactions and Adjustments)*. Concerns were raised as to whether it was too late for rain to help.

By mid-April, the Kansas Wheat Board reported that 76% of the state's winter wheat was in poor condition. By late April, following a further estimate that 48% of the winter wheat crop had been *lost*, the Kansas congressional delegation pressed the Bush Administration for action (*Chicago Tribune* April 23). It was further reported that 30% of the winter wheat fields would be abandoned—they were not worth harvesting.

The USDA took an approach of cautious optimism and announced on April 27 that the total U.S. wheat harvest would be up considerably from the 1.81 billion bushels harvested in 1988. They expected the total 1989 wheat crop to be 2.1 billion bushels, yet noted that even this amount would not be sufficient to rebuild depleted stockpiles (Associated Press April 27).

In general, by late spring the federal government was optimistic about the U.S. wheat picture, notwithstanding the problems in Kansas, Oklahoma, and Texas (Reuters May 12). The USDA announced on May 11 that the combined yields for Kansas, Nebraska, Oklahoma, and Texas wheat production would be down 30%, reflecting the dry conditions in those states (*Chicago Tribune* May 11). The situation was particularly severe in Kansas, where winter wheat production was predicted to drop by 37%. Total national hard red winter wheat production for 1989 was expected to drop 21% below the 1988 average, but the USDA calculated that overall winter wheat losses would be moderated by expected high yields of soft red winter wheat in the Midwest, for a decline in the total U.S. winter wheat production level of only 8% (*Chicago Tribune* May 12). Overall winter wheat losses were estimated to be $800 million (*McGraw News* May 11).

The USDA projected in May that U.S. wheat exports would decline by 300 million bushels below prior-year averages for 1989. Yet, the agency announced on May 15 that "U.S. wheat production will meet world desires and the losses of the High Plains would not essentially affect our production" (*U.S. News and World Report* May 15). As of June 1, 1989, however, the USDA forecast a meager 549 million bushels for wheat stocks, down 1.3 billion bushels from June 1, 1988 and down 1.82 billion bushels from June 1, 1987.

Problems with the 1989 spring wheat crop in the northern Plains also emerged as spring progressed. Drought damages led the governor of North Dakota to declare the state a disaster area. In mid-July, North Dakota estimated crop losses at $540 million, chiefly due to spring wheat losses (Reuters July 17). In its August 1 crop report, the USDA estimated that U.S. spring wheat crop production would be 12% lower than its July 1 estimate.

During this "off-season" drought, winter wheat farmers suffering losses waited with concern as Congress debated when and how much aid to extend to them. By mid-July, the USDA had shifted its views from "relief only for the winter wheat farmers" to considering expansion of drought relief to farmers of many other crops (Associated Press July 12).

*Livestock Producers*

The second group of individuals severely affected by drought during 1989 were livestock producers with herds in Iowa, Missouri, Nebraska, Kansas, Oklahoma, Texas, and New Mexico. The winter and spring precipitation levels in these states were well below normal. Spring 1989 ranked as the second driest since 1895 in Nebraska and New Mexico, fifth driest in Colorado, ninth driest in Iowa, 11th driest in Kansas, 15th driest in Missouri, and 18th driest in Oklahoma and Texas (Heim 1989).

For livestock producers, the impacts of the drought were two-fold. First, continuing drought had greatly reduced the quality of grass on pastures and rangelands. Second, 1988 drought losses in the Corn Belt greatly increased the prices of corn and soybean meal used for livestock feeding supplements (*News-Gazette* May 11). Thus, the impact of the *1988* drought on livestock producers became most serious during the spring of *1989*.

Assessments in May 1989 revealed that U.S. food prices would rise 5.5-6% due to the continued drought, principally from heightened costs of feeding livestock. This food-price increase was much higher than the USDA's November 1988 forecast of a 1989 rise of 3-5%. Agriculture Secretary Clayton Yuetter announced on June 30 that food prices would probably increase 7%, reflecting the large underestimation made previously. It was further announced in May 1989 that the drought of 1988-89 would produce long-term difficulties for the U.S. beef industry because of poor pasture lands and increased feed prices (*News-Gazette* May 11). The continuing problems of U.S. livestock producers were addressed in Congress beginning in early April, and a series of relief actions were taken by the federal government (see section on *Governmental Reactions and Adjustments*).

Continued drought in the summer of 1989, with higher grain prices both from the 1988 and 1989 drought seasons, also created a widescale liquidation of the national hog herd (*Farm Week* July 3, 1989), leading to a cut in pork production in 1990.

## Corn Belt Impacts

In the Midwest, farmers who suffered extreme drought in 1988 also experienced delayed impacts during the winter and spring of 1989. The USDA announced in early March 1989 that most Corn Belt farmers would participate in the set-aside program, leaving 10% of all lands fallow. In addition, many farmers could not afford 1989 planting costs because of financial losses in 1988 (*News-Gazette* March 5a). It was further announced in early March that the 1988 drought had reduced the effectiveness of pre-emergence herbicides, and that Corn Belt farmers would have to use greater amounts of expensive preplanting herbicides to control weeds in 1989. (*News-Gazette* March 5c).

The situation facing Illinois farmers typified the 1989 drought in the Corn Belt. Crop yields in 1988 had been reduced by 30-40%, but as shown in Figure 3.8, corn and soybean prices had increased dramatically during 1988 (*News-Gazette* March 5). Although this was a positive outcome for producers with corn and soybean stocks to sell, livestock and hog production costs had risen as a result of these price increases. However, the USDA believed that because the amount of set-aside land had been cut back for 1989 (causing more acreage to be put into production), grain prices would go down. In addition to feed and grain affordability problems, farmers were troubled over the effects of aflatoxin, a toxic mold present in the 1988 drought-damaged corn stored for 1989 feeding.

Forecasts for an end to the drought were common in 1989, even though there were indications of continued dry weather, such as the fact that corn and soybean plantings in Illinois, Iowa, and Missouri were behind average by May 1, 1989. Chief Meteorologist Norton Strommen of the USDA announced on April 26 that he was "optimistic" about prospects for all U.S. crops planted in the spring, because drought-related weather patterns were not prevalent (Reuters April 26). The general expectation of the USDA was that acreage planted to corn in 1989 would increase because of government incentives. However, soil moisture on May 1 was reported low over 48% of Illinois, 50% of Missouri, and 68% of Iowa (*News-Gazette* May 2), and a NOAA drought analysis indicated serious soil-moisture problems in the western Corn Belt and High Plains (Climate Analysis Center 1989). Additionally, excessive spring wetness in Indiana and Ohio reduced corn planting to about 2 million acres below the 73 million acres expected (*Farm Week* June 26). At the end of August, Illinois corn and soybean farmers, who were facing their second year of drought, decried the optimistic spring weather forecasts issued by both government and private meteorologists (*Farm Week* August 21a).

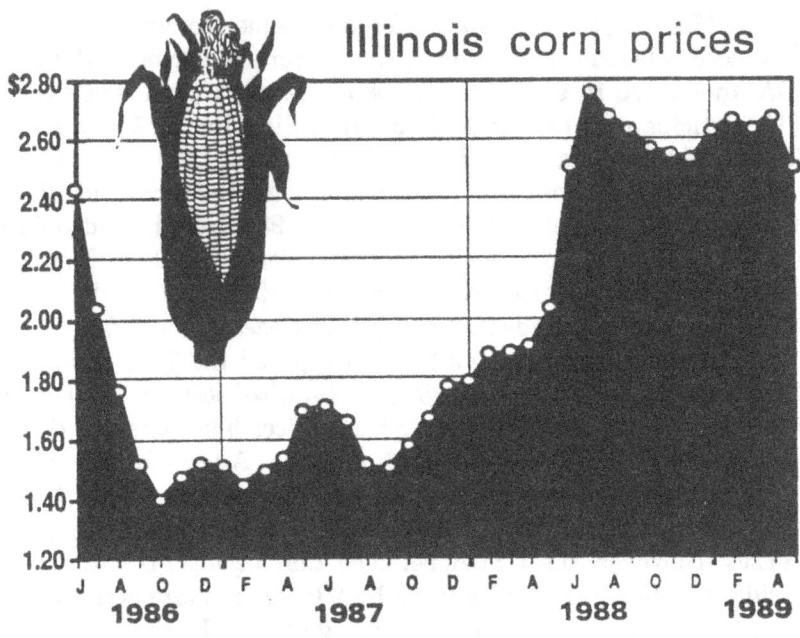

**Illinois corn prices**

Figure 3.8
Behavior of Illinois corn prices (*Source: Farm Week*, June 10, 1989)

On May 7, the USDA indicated that the 1988 drought had put between 10,000 and 15,000 of the nation's 550,000 commercial farmers out of business (*Los Angeles Times* May 7, 1989). A second dry year in 1989 would force an even larger number of farmers out of business. This situation was reflected in the demand for farm loans during the spring of 1989, up 15% over that of 1988 (UPI May 16d).

By the end of May 1989, a considerable climate dichotomy existed in the Midwest. In the eastern Corn Belt (Indiana and Ohio), the spring weather had been unusually wet and cold, which delayed planting (*Christian Science Monitor* May 17). This delay increased farm requests for short-season hybrids, and supplies of this type of seed were not adequate to meet the demand (UPI June 16). Many farmers shifted to soybeans. However, Ohio expected its 1989 yields of corn and soybeans to be 30% below average (UPI June 5b). On the west end of the Corn Belt—northwestern Illinois, large parts of Iowa, northern Missouri, and southern Minnesota—the opposite condition existed (*Farm Week* June 5, June 19). There, soil moisture was rated inadequate as of June 30 (*Farm Week* July 26a).

Herbicides applied to soybeans in 1988 were retained in the soil due to deficient 1988-89 precipitation, and chemical damage began to appear by the end of June 1989 in fields where corn was planted (*Farm Week* July 3c). This problem escalated in Iowa, where a group of farmers filed a class-action suit against three firms producing the herbicides (American Cyanamid, Elanco Products Company, and MFC Corporation). The chemical firms collectively agreed to handle farmers damage claims, but only after company officials performed in-field assessments (*Farm Week* July 24b).

The first two weeks of July in the Corn Belt and the High Plains produced frequent above-normal temperatures and very little precipitation. Concern spread rapidly because this was the period of corn tasseling and pollination, when hot, dry conditions can be very damaging. The USDA's July 1 estimates of corn production and surpluses fell sharply below earlier estimates. The fall 1988 surplus was 4.25 billion bushels of corn and 302 billion bushels of soybeans; but by July 1, estimates of 1989 fall surpluses were down to 2 billion bushels of corn and 125 billion bushels of soy beans (*Farm Week* July 10a).

By July, the weather stresses in the Corn Belt and High Plains had become national news, creating awareness that damage to wheat and corn crops was occurring for the second year in a row (Reuters July 12). But, the Corn Belt weather changed dramatically during the week of July 16-22, with widespread rains of one to three inches (Midwestern Climate Center 1989). Corn crops in western and northern Illinois, large parts of Iowa, and parts of Missouri and Minnesota were not productive, but were saved from total loss (*Farm Week* July 24a). At this point, the USDA announced for the first time that it would support a relief aid program for U.S. corn growers suffering from 1989 drought losses (*News-Gazette* July 25a).

Concerns then shifted to soybeans in the Corn Belt, which depend heavily on August rainfall for good yields. The outlook for adequate precipitation and moderate temperatures was good in Illinois, according to agricultural experts.

Early August crop assessments were performed by private firms and the USDA. The USDA August 1 crop report estimated the 1989 corn crop had decreased 7.35 billion bushels below the estimated yields calculated by most private firms (*Farm Week* August 14b). The expected U.S. surplus of corn in the fall of 1989 was 1.68 billion bushels, reduced from the 1.83 billion bushels estimated in July. Bean production was similarly downgraded in August, with a reduction in expected yield from 33 to 32.3 bushels per acre. By late August, 47% of the nation's soybean crop in 19 states was judged as fair to very poor (*Farm Week* August 21a). The drop in predicted yields by late August is depicted in Figure 3.9. Farmers in Illinois and Iowa were seeking governmental relief (*Farm Week* August 21b).

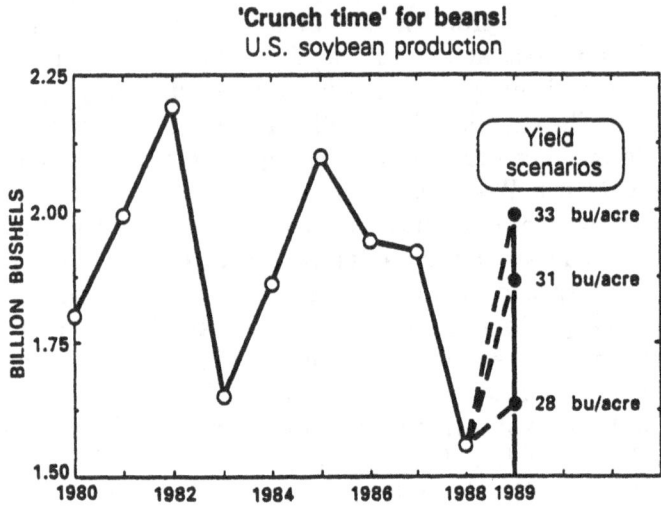

**Figure 3.9**
Fluctuation in U.S soybean production and yield scenarios developed in August 1989 for 1989 crop (*Source: Farm Week*, August 21, 1989)

*Grain Market Effects:*
*The Year of Hypersensitive "Weather Markets"*

One of the most fascinating effects of the 1987-89 drought was its unpredictable impact on the grain markets, especially in 1989. During this second year of problems, the markets became hypersensitive, swinging one way and then the other as forecasts, rumors, and reports of damage made their way to producers and traders. The 1988 drought had produced swings in grain prices beginning in June, but the 1989 oscillations were even more dramatic, and sometimes illogical.

A delayed reaction to the agricultural damages from the 1988 drought appeared in January 1989, when the USDA announced that the 1988 yield loss was actually less than earlier accounts (*News-Gazette* January 10). In November 1988, U.S. corn production was estimated at 4.9 billion bushels, but the actual figures in January indicated it was 5.1 billion bushels, causing grain prices to drop across the U.S. (*News-Gazette* January 17). In mid-April, prices of corn, oats, and soybeans rebounded dramatically due to reported increasing drought conditions in the western Corn Belt (Reuters April 17).

Prices of corn and soybeans fell on April 30, when widespread rains fell over the Corn Belt. One day later, prices of grain and soybeans rose considerably following the issuance of the 90-day weather outlook by the National Weather Service, calling for dryness in the western half of the Corn Belt. Experts reported that a "volatile weather market" had continued into 1989 (AP May 1).

The 1989 markets were especially weather-sensitive because, unlike prior years (including 1988), buyers in 1989 were not cushioned by large grain carry-overs, and another shortfall would cause major market price increases (*Financial Times* May 19; *Farm Week* July 10). Figure 3.8 illustrates the erratic behavior of Illinois corn prices over the last three years and reflects the drought's effect on market sensitivity.

The fluctuating market continued into late June. Soybean prices fell because weather forecasts indicated the likelihood of future precipitation (*Wall Street Journal* June 23). A few days later, prices for corn and soybeans rose suddenly following the issuance of a NOAA forecast for a hot and dry July. As the predicted weather pattern continued, prices edged up again, and some buyers were reported to have hedged against two consecutive droughts in the Corn Belt by hoarding grain inventories (*Farm Week* July 10a).

During July, a controversy arose at the Chicago Board of Trade (CBOT) concerning the hoarding activities of a major soybean market player, Ferruzi. CBOT ordered traders to reduce holdings in their soybean futures contracts, causing wild surges in soybean prices during July (*News-Gazette* July 25b). Ferruzi was forced to sell its contracts for the delivery of 223 million bushels of soybeans, which it was attempting to hold because of the drought-reduced 1988 crop by claiming it required sufficient soybean supplies to meet its long-term needs (*News-Gazette* August 16). Soybean prices further decreased in late July with issuance of wet weather forecasts for August and the record Brazilian soybean crop of 1988-89 (*Farm Week* July 31a). Market hypersensitivity continued through the remainder of the summer.

*Agribusiness*

Most U.S. agribusinesses had been affected in some way by the 1988 drought. Many large agribusinesses developed corporate drought task forces during the summer of 1988, and these in turn produced new plans for marketing, production, and sales in 1989. One major company formed its own weather information group. Increased purchases of weather information and climate prediction services were noted by private meteorologists in 1988 and 1989.

Several midwestern seed companies experienced severe losses of seed crops in 1988 (*News-Gazette* March 4). Thus, during the 1988-89 winter season, they contracted to buy seed crops grown in Texas, Florida, and South America. Several firms also had depleted their over stocks in 1989, when demand was high for corn and soybean seed. Additionally, there were insufficient stocks of short-season hybrid seed for Indiana and Ohio farmers facing a short growing season.

The farm machinery manufacturing industry announced in March 1989 that it was enthusiastic about future sales (*News-Gazette* March 5d). This optimism developed because the 1988 drought had led to low carryover stocks, higher commodity prices, an expected *increase* in farm income due to higher prices and government aid, and more land in production in 1989 because of government incentives. (See Chapter 3, Section I).

One agribusiness industry seriously affected by the drought was crop insurance. The Farm Relief Act of 1988 required any farmer accepting relief payments to purchase all-weather peril insurance during 1989 and 1990. Thus, insurance sales were much above expected levels, nearly doubling the 1988 sales level.

The rain insurance business particularly suffered from the 1988 drought. On April 20, 1989, the Chubb Insurance Group paid out $19.2 million via 2,838 checks in 10 states as part of a settlement with farmers carrying rain insurance (*News-Gazette* April 21). Chubb had already paid out $37.4 million to farmers for insured losses. In turn, the Chubb Group and 8,800 farmers jointly sued Good Weather International Corporation, the associated company that designed and marketed the 1988 rain insurance coverage. The litigants accused Good Weather of selling more policies than Chubb had authorized (UPI July 13). Chubb claimed that Good Weather violated its contract by selling too many crop insurance policies and that Good Weather owed Chubb at least $75 million because of the payoff costs and lawyer fees. In turn, Good Weather denied all claims by the farmers and Chubb. By mid-August, the farmers and Chubb endorsed an out-of-court settlement of $8 million from Good Weather International Corporation, with $3.7 million going to Chubb Corporation (*News-Gazette* August 15).

Financial institutions, including farm banks, were affected, especially by the bankruptcy of some 10,000 farms in 1988 and 1989, some of which resulted from the drought. However, most banks in the Midwest reported no major losses in 1988, and farm loans increased in the spring of 1989 by 15% over 1988 as producers geared up (UPI June 1).

## Governmental Reactions and Adjustments
## to Agricultural Problems

State and federal responses to the agricultural drought problems in 1989 focused on two groups: (1) the High Plains winter wheat farmers and (2) crop and livestock producers concentrated in the western Corn Belt and High Plains.

Drought effects in the Corn Belt first attracted national government response when Senator Bond of Missouri announced in early April that Congress would need to provide drought relief in 1989 (UPI April 4). Winter wheat problems in Kansas became national news soon afterward, and members of Congress began drafting disaster relief legislation for wheat farmers (Reuters April 10).

The Bush Administration and the USDA initially disagreed with Congress about the need for relief, arguing that the severity of the drought was still uncertain and that it was too early to react (*Los Angeles Times* April 10). The USDA's chief meteorologist said in March that it was "premature" to assess 1989 drought impacts and that the weather patterns of 1989 were different than those of 1988 (*Los Angeles Times* April 10). The major initial debate became whether it was "too late for rain to help the crop." Essentially, the USDA was saying that it was not too late, whereas the wheat farmers and their congressional representatives were saying that it was.

The second point of contention was whether to provide relief to wheat farmers or allow national crop insurance to pay those who had purchased it (Reuters April 10). USDA disaster payments had been curtailed in 1982 to encourage the purchase of crop weather insurance. Yet by 1988, only 29% of cropland was covered, raising criticisms of the national insurance program (Reuters April 10). The 1988 Drought Assistance Act, with nearly $4 billion in relief payments, had further eroded interest in insurance and contradicted the 1982 policy of weaning farmers from repetitive disaster aid. The basic principle held by Wheat Belt members of Congress was that "anyone hurt by this natural disaster deserves assistance," creating an insurance-versus-relief controversy.

By April 23, the Kansas congressional delegates were openly seeking special relief, despite the "wait-and-see" position of the Bush Administration (*Chicago Tribune* April 23). At the same time, requests for relief aimed at the livestock industry were made. The governor of Iowa requested two forms of assistance from USDA Secretary Yuetter: (1) open up grazing land in the set-aside program and in the Conservation Reserve Program (CRP), and (2) provide livestock producers cheaper feed through government subsidy or purchase from government grain stocks (UPI April 25). In Washington, the livestock problems were seen as easier to solve than the enormous wheat problem, and the administration acted quickly—the USDA announced on the next day that if a county had lost 40% or more of its pastures, it qualified to cut the hay or graze livestock on reserve land (Reuters April 26).

On April 27, the USDA announced that farmers could also purchase surplus grain from the Commodity Credit Corporation at half price. Secretary Yuetter also established a USDA Drought Task Force to monitor the drought and to consider relief payments for 1989 crop losses (AP April 27). In early May, wheat belt legislators sought a drought relief plan much like that passed in 1988 (UPI May 3).

Hearings about crop losses were held by members of Congress in Texas and Kansas. Senator Robert Dole of Kansas estimated that the cost of a 1989 agricultural drought relief plan would be approximately $350 million (UPI May 4). Dole further indicated that, as in 1988, this cost would be offset by much smaller crop subsidy payments because of lower wheat yields. The USDA responded on the same day that they were "watching the weather conditions" (UPI May 4).

On May 11, Agriculture Secretary Yuetter announced that farm drought relief, like 1988, "may be necessary" in the winter wheat areas (*Chicago Tribune* May 11). The estimated wheat loss in the Great Plains was 21%, with Kansas suffering crop losses of 37%. NOAA meteorologists predicted on the same day that the drought was expected to continue in Kansas and the High Plains (*News-Gazette* May 11).

The governors of Kansas, South Dakota, and North Dakota joined the debate in mid-May by asking Congress for help in passing a drought relief bill (UPI May 16b). On May 22, the Governor of Wisconsin convened a drought task force because of severe damage to the alfalfa crop (UPI May 24), and the Governor of Illinois assembled his drought task force on May 24 (*Chicago Tribune* May 24). At the same time, Iowa established an emergency water plan for livestock producers (UPI May 22b).

The House Agriculture Committee approved a drought relief bill on May 26, despite the disapproval of the USDA. The House passed the $1 billion aid bill on June 27 (AP June 27), disregarding protests by the Bush Administration. Because an aid package seemed inevitable, Secretary Yuetter sought assurances from Congress that, after 1989, only crop insurance would be used as disaster relief, essentially delaying the eventual debate over crop insurance versus disaster relief.

Wheat Belt senators were also trying to extend the 1988 Farm Relief Bill. Senator Bond of Missouri asked the USDA for aid beyond the emergency grazing and haying provisions. On June 2, Senator Patrick Leahy, Chairman of the Senate Agricultural Committee, indicated that the Senate would delay legislation for farm relief (AP June 2). An important senate debate concerned whether the 1989 drought relief action should include all crops or just winter wheat. Senators Leahy and Lugar (the ranking members of the Agricultural Committee) finally resolved their differences and agreed to act on drought relief on July 19 (AP June 14). On June 28, President Bush announced that he would oppose a drought relief bill if it contained too many crops under its "umbrella" (AP June 28).

By early July, with increasing impacts, the USDA's position on the continuing drought, and on the overall impact of the drought of 1988, was still ambiguous. Secretary Yuetter analyzed drought-depleted stocks and concluded that the U.S. had a "comfortable supply" (Bureau of National Affairs July 5). Based on its July 1 crop estimates, the USDA announced two other significant trends. First, the expected 1989 corn crop yield of 7.85 billion bushels (as estimated in March) was now reduced to 7.45 billion bushels (Reuters July 12). However, no decrease in soybean production was expected. Second, the USDA indicated for the first time that the government might have to expand its assistance to farmers suffering from losses of several other 1989 crops, repeating that it was "an unusual weather year" (Figure 3.10) and noting the need also to aid farmers who had experienced too much rain (*Farm Week* July 17a). By these statements, the USDA had shifted from a wait-and-see, wheat-only relief policy to one that included other major crops, such as rice and cotton (AP July 12).

The Senate seriously took up the issue of agricultural relief in mid-July. Its Agricultural Committee embarked on a heated, politically aligned debate over the relief bill. Republicans wished to minimize the extent of emergency aid and Democrats wished to expand it (AP July 20). By July 25, the committee approved a 1989 drought relief bill, with 10 Democrats voting in favor of and nine Republicans voting against the measure (UPI July 25), which provided $1 billion in farm aid.

Senator Leahy, Chairman of the Agricultural Committee, indicated that he wanted the relief bill agreed upon by both the House and Senate by August 4, the day before Congress adjourned for its one-month summer vacation. The Bush Administration threatened to veto any act that exceeded $1 billion. Senator Dole announced that he would scuttle any bill over $1 billion and that the Democratic version, amounting to $955 million, was too high. The Senate Agricultural

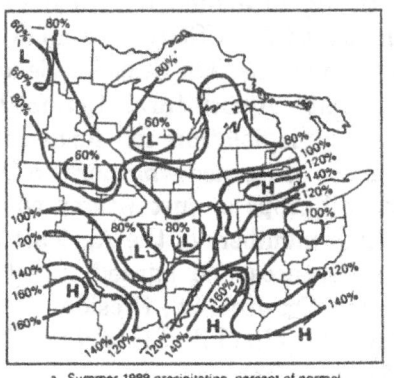

a. Summer 1989 precipitation, percent of normal

b. Summer 1989 temperatures (°F), departures from normal

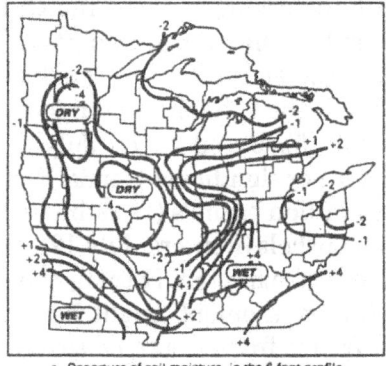

c. Departure of soil moisture, in the 6-foot profile,
from normal at end of August 1989 (values in inches)

**Figure 3.10**
Drought conditions during the summer of 1989 (*Source*: Midwestern
Climate Center)

Committee met on July 25, at which time Republican members agreed
to include soybean growers, but debates continued over the amounts
of coverage for each crop (*News-Gazette* July 26).

Agricultural interests around the nation became highly sensitized
to the political discord, which was seen as jeopardizing the chances
for a timely relief bill (*Farm Week* July 31b). The American Farm
Bureau supported the Democrat's version of the bill, and asserted that
relief should be provided to producers of all affected crops.

In an effort to complete its work before adjournment, the Senate
Agricultural Committee approved a 1989 Drought Relief Act, amounting to $885 million, on August 2 (*News-Gazette* August 3). On August
4, one day before the congressional vacation, the House and Senate
agreed to a mixed relief package of $900 million after agreeing that
relief would go to producers of all crops, but that payments would be
set at varying levels (*News-Gazette* August 5). Secretary Yuetter indi-

cated that the $900 million level of relief was acceptable. The bill that the House and Senate agreed upon provided the greatest support to farmers who had participated in federal farm programs and those who had federal crop insurance in 1989. The bill allowed for losses due to both drought and extremely wet conditions during 1989.

President Bush signed the 1989 Farm Relief Act into law on August 15, almost exactly a year after President Reagan signed the 1988 Farm Relief Act. However, the 1989 relief act provided only $1 billion for assistance, or 25% of the $4 billion provided in 1988.

## Summary

Uncertainty over crop yields and appropriate actions, political debate, and continued poor assessment of drought conditions characterized government response to the 1989 drought. Federal policy was influenced by several issues. The first was the likelihood of two back-to-back severe drought years. During the winter of 1988-89, USDA meteorologists stated that the chance for severe drought in 1989 was slim. Though the nation as a whole was not as dry in 1989 as in 1988, some areas did experience severe drought, and it is reasonable to conclude that the U.S. did, in fact, experience two consecutive drought years.

Another issue centered around the use of drought relief versus crop insurance for easing losses; many policy makers felt that ad hoc relief undermines the logic of insurance. Yet, earlier agreements to encourage insurance, rather than disaster relief, had broken down, essentially because farmers had not widely used the insurance program. It should be noted that the relief/insurance debate is also a central issue in federal programs that address other hazards (e.g., floods and earthquakes), and deserves more considered study.

A fourth issue was how to provide relief payments for crop losses in 1989. Some wanted to write a new relief act for 1989, and others wanted to extend the 1988 Farm Relief Act while waiting to see the extent of losses after the 1989 harvest. A central question focused on what crops to cover in the 1989 Relief Act—initially wheat only, then just "program crops" (those receiving subsidies), and finally all damaged crops (*Farm Week* July 3).

Government response to these issues in 1989 revealed a continuing problem: monitoring and impact assessments of the 1989 drought were inadequate, and provided little guidance for a rational relief program. If anything, there was greater uncertainty over drought extent, continuance, and potential impacts in 1989 than in 1988. Meteorologists expressed uncertainty over whether the drought would continue, redevelop, or terminate in 1989, and if present, how severe it was. These issues, coupled with constant uncertainty about the drought's effects on crop production, caused continued conflict among congressional lawmakers and between Congress and the administration.

Agricultural policy in 1989 revealed a shift in the administration's attitudes. During the spring of 1989, the views expressed largely through the USDA were of cautious optimism that downplayed the potential impacts of the drought, particularly to summer crops. As conditions worsened during May, June, and July, however, the position of the USDA shifted from relative optimism to realism, with admissions in July that the nation not only had drought damage to the winter wheat crops of the High Plains, but also faced losses of corn and soybeans in the western Corn Belt and losses of spring wheat in the northern Great Plains. The initial optimistic outlook could be credited to several factors, including a desire to present a bright outlook to U.S. grain exporters and foreign importers, and to a general optimism that a multi-year drought was not likely.

The drought problems of 1989 also raised issues concerning future agricultural policies. The 1990 farm bill will touch on many issues made more controversial by the drought, such as the debate over the roles of crop insurance and disaster relief.

The drought may also have an effect on corn set-aside levels, as farmers put more land into production to compensate for drought losses. On July 3, 1989, the USDA had forecast that the amount of lands set aside in 1990 would be greater than 12.5% of lands in corn, but this had shifted by late July (with larger crop loss estimates) to less than 12.5%. (It was 10% in 1988.)

A major national policy concern raised by the droughts of 1988 and 1989, and important to future drought response, is the national grain surplus and the role it plays in U.S. commerce and loss mitigation, and global food supplies. The USDA's 1988-89 grain production and surpluses policies, including export targets in 1989, were based on a nondrought scenario for 1989. As noted, this scenario stemmed essentially from optimism, rather than climatological history, and was based on the expectation that the weather would return the U.S. to "normal agricultural production in 1989."

Table 3.7 shows how those expectations for production, and in turn surpluses, had to be revised downward as 1989 progressed. Several events, some which were foreseeable, overrode the USDA's optimistic expectations for 1989 crop yields.

## Table 3.7
## USDA Estimates of 1989 Corn Crop
## and Available Surpluses

|  | Fall 1988 | March 1 1989 | July 1 1989 | August 1 1989 |
|---|---|---|---|---|
| Production (billions bushels) | 7.85 | 7.45 | 7.35 |  |
| Acres (millions acres) | 73.3 | 72.7 | 65.15 |  |
| Yield (bushels/acre) | 114 | 113 | 112.8 |  |
| Stocks (billions of bushels) | 4.25 | 2.0[a] | 2.83[a] | 1.68[a] |

[a]Expected by Fall of 1989
Note: the U.S. annually utilizes 7 billion bushels of corn.

The obvious question raised by the surpluses, and the related issue of linking relief to "savings" in subsidies (as debated during the drafting of the 1989 relief legislation), is simply whether economically driven surpluses and politically motivated support programs can be counted upon as reliable drought adjustments. The USDA, farmers, and agribusiness are diligently working to *reduce* grain surpluses (preferably by increased foreign sales). Moreover, Congress, in response to the diminishing political power of the "farm block" and federal deficit reduction goals and mandates (including those mandated by Gramm-Rudman), is seeking to eliminate grain subsidies over the next several years. As a result, a *de facto* drought mitigation policy is targeted for deletion, while no specific drought policies have been proposed to mitigate the obvious vulnerability that will then emerge.

### References

Associated Press, April 27, 1989: "Yuetter Renews Last Year's Drought Assistance."

——, May 1, 1989: "Gloomy Weather Forecasts Grains Higher."

——, May 11, 1989: "Researchers Stymied on Pig Deaths. Could be Drought Related."

——, June 2, 1989: "Leahy Says Congress Will Move Slowly on Drought Relief."

——, June 3, 1989: "Drought May Be in Remission, But Food Prices Still Headed Higher."

———, June 14, 1989: "Rural Development Passes Committee with Accord on Drought Relief."

———, June 19, 1989: "Corn Closes Limit Up on Weather, Load Extension."

———, June 27, 1989: "House Passes Drought Relief for Farmers."

———, June 28, 1989: "House Drought Relief Bill for Farmers Faces White House Opposition."

———, June 29, 1989: "Cherry Output Up, But Still Suffering from Last Year's Drought."

———, July 11a, 1989: "Jury Picked for Agent that Sold Drought Insurance."

———, July 11b, 1989: "Midwest Sizzles."

———, July 12a, 1989: "Yuetter Appears to Open Door for Expanding Drought Relief."

———, July 12b, 1989: "Heatwave Continues."

———, July 20, 1989: "Senate Committee Deadlocked on Drought Relief."

Bureau of National Affairs, July 5, 1989: "Drought Depleted Stocks Leave U.S. Comfortable Supply."

*Champaign-Urbana News-Gazette,* January 1, 1989: "Drought Dilemma Dominates 1988."

———, January 5, 1989: "Drought, Greenhouse May Be Unrelated."

———, January 10, 1989: "Crop Report Raises 1988 Output."

———, January 17, 1989: "Grain Prices Drop Sharply Despite Poor Crop Report."

———, January 21a, 1989: "ADM Buys 8% of Illinois Central as Investment."

———, January 21b, 1989: "Despite Drought, Missouri Ducks and Hunters Fared Well."

———, January 21c, 1989: "Irrigation that Saved Crops also Depleted Well Supplies."

———, February 2, 1989: "House Ag Chairman: Drought Long Way from Over."

———, February 6, 1989: "Moisture or Lack of it, Remains a Cause of Concern."

———, February 11, 1989: "Expert: Soil Moisture Likely to Recharge by Spring."

———, February 20, 1989: "Top Meteorologist: Snow is Making Drought Unlikely."

———, February 26, 1989: "California Drought."

———, March 1, 1989: "Mississippi Swells as Rain Level Recovers."

———, March 3, 1989: "Weather Extremes Forecast for Spring."

———, March 4, 1989: "Caution Becomes Bi-word in Assessing State Farming."

———, March 5a, 1989: "Farm Managers Expect Increased Role."

———, March 5b, 1989: "Congressmen: Few Major Changes in Farm Policy are Likely in 1989."

———, March 5c, 1989: "Drought May Spur Change in Herbicides."

———, March 5d, 1989: "Drought Still has Effects: Higher Prices for Seed Hamper 1989 Activities."

———, March 5f, 1989: "Experts: Soil Moisture in State Makes Rerun of Drought Unlikely."

———, April 21, 1989: "CHUBB Mails Drought Insurance Checks to Farmers."

———, April 30, 1989: "Crop Prices Fall Along with Rain in the Midwest."

——, May 2, 1989: "Illinois Farmers Find Dry Subsoil with About Half of Corn Crop In."

——, May 4, 1989: "Water Utility to Seek Hike in Rates and Fees."

——, May 11, 1989: "Cost of Filling Up Your Car and Your Table Increasing."

——, May 11, 1989: "It's 1988 Drought Again for Kansas"

——, May 26, 1989: "Low River Could Trap Barges."

——, May 29, 1989: "Ag Expert Says Low Mississippi Could Slow Grain Barges this Fall."

——, June 10, 1989: "Drought Creates Problems that Trees Can Leave Behind."

——, June 20, 1989: "Plains See Worst Wind Erosion Since 1954."

——, July 11, 1989: "Two Die as More than 11,000 Fight Western Fires."

——, July 12, 1989: "Rain Helps Fire Fighters Gain Ground."

——, July 15, 1989: "People Wonder if Extreme Weather Stems from Greenhouse Effect."

——, July 25a, 1989: "Drought Relief Bill will Include Corn, Iowa Governor Told."

——, July 25b, 1989: "Ferruzi Wants Study of CBOT Operations."

——, July 26, 1989: "Drought Bill Sparks Partisan Feud."

——, August 2, 1989: "Fire Leaps River, Troops Move in to Help on Lines."

——, August 3, 1989: "Senate Okays Drought Rural Development Bills."

——, August 4a, 1989: "Cool Weather Aids Fire Fight in Idaho."

——, August 4b, 1989: "U.S. Shipped High-quality Grain Despite Drought."

——, August 4c, 1989: "Corn and Soybean Potential Looks Good."

——, August 5, 1989: "Congress Earmarks $900 Million to Pay for Crop Losses."

——, August 6, 1989: "Lightning Forecast Worries Tired Firefighters in West."

——, August 13, 1989: "Mild Weather Reduces Aflatoxin Occurrence in Corn."

——, August 15, 1989: "Farmers' Lawyer Endorses $4.93 Million Settlement."

——, August 16, 1989: "CBOT Won't Punish ADM Chief."

Changnon, S.A., "The 1988 Drought, Barges, and Diversion." *Bulletin of the American Meteorological Society*, Vol. 70: 1092-1104. 1989.

*Chicago Tribune*, April 16, 1989: "Bad Weather after Drought and a Wimpy Winter, Gardeners May be Bugged."

——, April 23, 1989: "Drought Rerun Guts Wheat: Kansas Faces Worst Harvest in 30 Years."

——, May 8, 1989: "Aided by 1988 Drought, Nation's Farmland Values Rise by 6%."

——, May 11, 1989: "Drought Aid Studied for Wheat Crop."

——, May 12, 1989: "Drought to Reduce Wheat Crop 8%."

——, May 24, 1989: "State Task Force on Drought."

——, May 29, 1989: "Odds Against '89 Drought in Illinois."

——, June 19, 1989: "Drought Persists in Illinois Despite Soil Moisture Gains."

——, June 28, 1989: "Weather Produces a Mosquito Boom."

——, June 28, 1989: "Drought Bill Gains, but Survival in Doubt."

——, July 8, 1989: "East is Storm; Southwest Bakes."

*Christian Science Monitor*, May 17, 1989: "Wet, Cold Weather Delays Planting."

——, June 2, 1989: "Rains Saturate Eastern U.S., but Drought Continues in West."

Climate Analysis Center. *Drought Advisory*. Washington, D.C.: National Oceanic and Atmospheric Administration. 1989.

*Farm Week*, May 29, 1989: "Markets Down-grade Drought Prospects."

——, June 5, 1989: "Crop Hopes Ride on More Rains."

——, June 19, 1989: "Summer Weather Prospects Eyed."

——, June 26a, 1989: "Can Feisty Crops Fend Off Foes"?

——, June 26b, 1989: "Fewer Corn Acres Seen."

——, July 3, 1989: "The Politics of Drought, 1989."

——, July 10a, 1989: "Markets Swing on Rain Chances."

——, July 10b, 1989: "Stress Building on Crops."

——, July 10c, 1989: "Herbicide Carryover Goes to Iowa Courts."

——, July 17a, 1989: "Yuetter Addresses Wide Range of Farm Issues."

——, July 17b, 1989: "Fewer Acres Help Offset Price Plunge in Soybeans."

——, July 24a, 1989: "Rains Fend Off Crop Disaster."

——, July 24b, 1989: "Chemical Firms Reach Pact."

——, July 24c, 1989: "Smaller Corn Set-aside Seen as 1989 Crop Prospects Cut."

——, July 24d, 1989: "Beef Outlook May Spur Feeder Bids."

——, July 31a, 1989: "Will Bean Price Fuse be Doused"?

——, July 31b, 1989: "Senate Divided Over Drought Aid."

——, August 7a, 1989: "Disaster Aid Passed: Bush Signature Seen."

——, August 7b, 1989: "USDA Weather Expert Sees Climate Knowledge Gap."

——, August 14a, 1989: "No Yield Records to Fall, but Crop Scout Satisfied."

——, August 14b, 1989: "Tighter Fit for Corn Users."

——, August 21a, 1989: "The Drought that Won't Die."

——, August 21b, 1989: "Drought Victims Wait for Assistance Signup."

*Financial Times*, May 19, 1989: "U.S. Cattlemen Suffer in Drought."

——, May 24, 1989: "U.S. Grain Traders Watch the Weather."

Heim, R. *United States Spring Climate in Historical Perspective*. Asheville, North Carolina: NCDC. 1989.

Illinois State Water Survey. *Water Status Report*. July 1989.

Interagency Drought Policy Committee. *The Drought of 1988*. Final Report of the President's Interagency Drought Policy Committee. Washington, D.C.: White House. 1988.

Lebham-Friedman, Inc. "Drought Dampens Economic Growth." May 8, 1989.

*Los Angeles Times*, April 10, 1989: "Kansas Loses 25% of Wheat to Drought."

——, April 14, 1989: "Farmers Prepare for Planting Amid Signs Drought is Still Plaguing Midwest and South."

——, May 7a, 1989: "Drought Producing a Bountiful Crop of Boom-to-Bust Tales."

——, May 7, 1989: "American's Heartland: The Exodus is Underway."

——, May 14, 1989: "Iowa Farmers Wait, Wonder if Another Drought Looms."

——, July 16, 1989: "Persistent Drought Plagues Midwest for Second Year."

McGraw Hill News, May 11, 1989: "Drought Outlook for Midwest is Grim, Experts Say."
Midwestern Climate Center. *Climate Impacts for July 1989.* Champaign, IL. 1989.
*New York Times,* May 16, 1989: "Diverted Jet Stream Creates Weird Weather Patterns."
*News Day,* May 2, 1989: "Discovery: Just a Drop in the Bucket."
Reuters News, May 24, 1989: "U.S. Department Extends Drought Aid to Ranchers."
———, April 10, 1989: "Emergency Drought Aid to Farmers may Become Permanent."
———, April 17, 1989: "Grain Prices Jump on Drought Worries."
———, April 26, 1989: "U.S. Offers Drought Aid to Livestock, Poultry Producers."
———, May 12, 1989: "Drought Taking its Toll on Wheat Farmers."
———, May 25, 1989: "Drought Bill Clears Panel Despite Administration Objections."
———, July 3, 1989: "Corn and Soybeans Rise on Dry Weather Forecast."
———, July 12, 1989: "Weather Takes Toll on Corn Crop for Second Year."
———, July 17, 1989: "Drought Plagues Upper Midwest."
*St. Louis Post Dispatch,* January 23, 1989: "If Spring Turns Dry."
*San Francisco Chronicle,* January 21, 1989: "Drought Still Grips Plains, Experts Say."
*Sports Illustrated,* March 13, 1989: "A Climate for Death."
UPI, April 4, 1989: "Bond Says More Drought Relief Possible."
———, April 13, 1989: "Indiana Farmers Get $16 Million in Drought Aid."
———, April 24, 1989: "1988 Drought Affecting Hay Prices."
———, April 25, 1989: "Branstad Asks for Livestock Drought Aid."
———, May 1, 1989: "Coast Guard, Shippers Guardedly Optimistic About a Drought."
———, May 1, 1989: "Northern Missouri Drought Continues."
———, May 3, 1989: "Lawmakers Unveil Drought Plan."
———, May 4, 1989: "Wheat-state Leaders Push for Drought Relief."
———, May 8a, 1989: "Drought Still Taking its Toll on Shrubs."
———, May 8b, 1989: "Rains Improve Crop Report, but Drought Continues."
———, May 8c, 1989: "Drought Aid Reaches $3.5 Billion."
———, May 8d, 1989: "1988 Drought Hurts 1989 Fruit Crops."
———, May 9, 1989: "Drought Aid Okayed for 158 Counties."
———, May 13, 1989: "Great Demand for Maine Potatoes: Last Year's Drought Increases Demand for Maine Spuds but Supplies Limited."
———, May 16a, 1989: "Counties Approved for Drought Aid Doubles."
———, May 16b, 1989: "Drought-State Governors Ask for Congressional Help."
———, May 16c, 1989: "Lightfoot Asks for More Drought Relief Measures."
———, May 16d, 1989: "Farm Borrowing Up Due to Last Year's Drought."
———, May 17, 1989: "A Repeat of Drought Possible in '89."
———, May 18, 1989: "USDA Meteorologist Says Rain Won't Cure Winter Wheat Crop."
———, May 22a, 1989: "Thompson to Reconvene Drought Task Force."

——, May 22b, 1989: "Drought Plan for Livestock Producers Announced."

——, May 30, 1989: "Yuetter Tours Drought Area."

——, June 1, 1989: "Fed Says Banks Did Well in Upper Midwest Despite Drought."

——, June 5a, 1989: "Drought Aid Ok'd for 582 Counties."

——, June 5b, 1989: "Wet Weather Changes Corn and Beans Decisions."

——, June 8, 1989: "Five States Get Half of 1988 Drought Aid."

——, June 12, 1989: "Bond Says More Drought Relief Needed."

——, June 13, 1989: "Drought Aid Coverage Area Expands."

——, June 16, 1989: "Wet Weather Costly to Ohio Farmers."

——, June 27, 1989: "Cochran Praises Drought Action, Criticizes Yuetter."

——, July 7a, 1989: "1988 Drought Aid Helped 790,000 Farmers."

——, July 7b, 1989: "Drought Continues in North-Central Illinois."

——, July 13, 1989: "Agent 'Ran Wild' on Drought Insurance, Chubb Says."

——, July 17, 1989: "Drought Aid Approved for 925 Counties."

——, July 23, 1989: "Hardships Continue in Drought-Stricken Northwest Missouri."

——, July 25, 1989: "Senate Committee Approves Drought Aid Measure."

——, July 27, 1989: "Scattered Storms and Heat."

*U.S. News and World Report*, May 15, 1989: "A Springtime Drought with a Silver Lining."

*Wall Street Journal*, June 23, 1989: "Grain Soybean Futures Take a Plunge as Fears of Another Drought Fade."

*Washington Post*, May 21, 1989: "Midwest Drought is Older but not Wider this Year: A Dry Spell Lingering in Pockets of Great Plains."

# 4

---

# Drought, Barges, and Diversion in the Mississippi Basin

## Introduction

One of the more interesting consequences of the drought was the impairment of midsummer barge movement caused by low flows on the Mississippi, Ohio, and Missouri rivers—the major rivers that drain most of the central United States. National attention was drawn to the difficulties by a controversial proposal: divert the waters from Lake Michigan into the Illinois and Mississippi rivers to increase their levels.

Together, these events provided valuable lessons for scientists and policy makers. They illustrated the use (and nonuse) of weather and climate information, the economic value of long-range climate predictions, and the need for faster reliable climate information for decision making (Changnon, et al. 1988). These events also foretold problems that a drier future climate could create in the humid eastern U.S., particularly if hypothesized $CO_2$-induced greenhouse effects are realized (Koellner 1988). Rapid response to the river blockages by the U.S. Army Corps of Engineers (USACE), and ensuing shifts by producers to alternative shipping modes, revealed the resilience of existing infrastructure and support systems and the value of redundancy in the transport of climate-sensitive commodities.

This case study characterizes the 1988 drought's creation of unusually low streamflows in late spring, the problems that ensued, the various responses proposed and employed, and the major winners and losers. Implications for future research, policy, and drought planning are identified.

## Background

"The Mississippi River Navigation System is entirely dependent on its abilities to transport commodities efficiently" (Koellner 1988), but this efficiency was greatly reduced in 1988. Barge movements were restricted in midsummer by low streamflows in the Ohio and Mississippi River channels south of the lock and dam systems on each river, especially south of Cairo, Illinois (Figure 4.1). This series of locks and dams controls the movement of water, helps prevent flooding, and sustains flows for waterborne transportation, power generation, irrigation, urban water supplies, and environmental quality.

South of these controlled-flow systems, river levels were unusually low by late May 1988, leading to increased sediment deposits. As a result, barge traffic was halted at several locations on both rivers over a four-week period, and traffic lessened throughout the summer. The major commodities hauled on barges (grains, petroleum, chemicals, and coal) were diverted to railroads and ships operating from Great Lakes ports and away from those operating out of New Orleans.

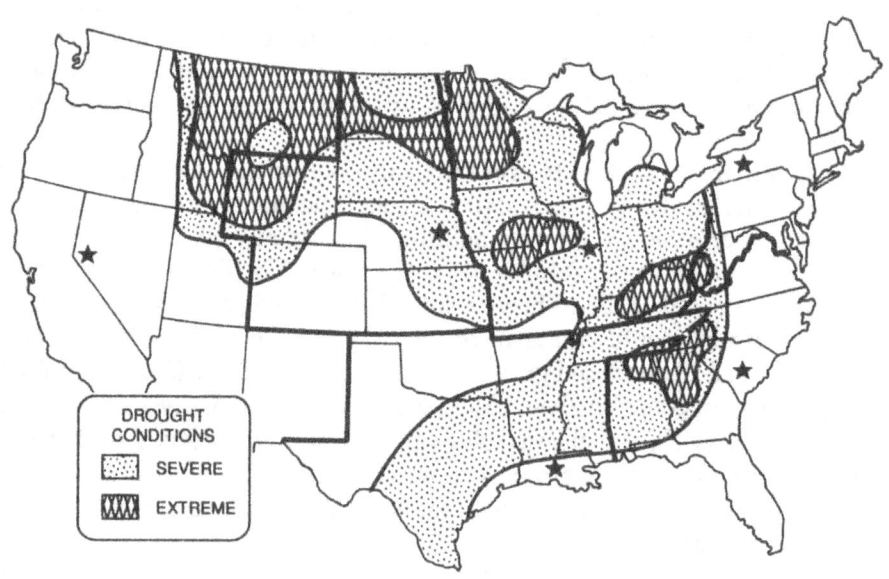

**Figure 4.1**
The areas of severe and extreme drought in the Mississippi River Basin on June 15, 1988 (based on the Palmer Drought Severity Index) (*Source:* Midwestern Climate Center)

More than 300 tow and barge companies operate on the Ohio, Mississippi, and Illinois river systems, and many river ports serve the barges. The annual revenue of the barge industry is approximately $1 billion per year (*Champaign-Urbana News-Gazette* August 1). The barge and tow industry carries 60% of all grain exported from the U.S., and 40% and 20% of all petroleum and coal, respectively, transported within the U.S. (American Waterways Operators 1988). Barge shipments typically transport 45% of the entire midwestern grain crop (*Chicago Sun Times* July 24). Thus, the barge industry is one of the nation's major conveyors of bulk commodities, and as such, is a key U.S. transportation industry.

## Antecedent Climate and Streamflow Conditions

Prevailing monthly temperatures and precipitations in 1987 for three regions—the Missouri Basin, the Upper Mississippi-Ohio Basin, and the Tennessee-Lower Mississippi Basin—are shown in Table 4.1. The 1987 sums reveal the preponderance of warm and dry conditions in all three basins, with few months rated as relatively cool and wet.

The warm and dry conditions during 1987 led to reduced stream-flows in the three major areas that drain into the river systems: the northern High Plains (North Dakota, South Dakota, Minnesota, Iowa, and Nebraska), the western Great Lakes (Illinois, Wisconsin, Indiana, Ohio, and Michigan), and the Southeast (Kentucky, Tennessee, West Virginia, North Carolina, South Carolina, Georgia, Alabama, and Mississippi). Data on average monthly flows of all gauged streams in these areas appear in Figure 4.2, expressed as percentage departures from the median discharge for 1951-80.

Flows in the northern High Plains (Figure 4.2a) and western Great Lakes (Figure 4.2b) were both consistently well-above median levels during 1986, but fell below median levels during the spring and summer of 1987. Heavy rains in August 1987 restored flows to near median levels. Above-average precipitation in November and December 1987 produced, with the normal lag time, slightly above median flows in both areas from December 1987 to February 1988. The average flow in the Southeast (Figure 4.2c) illustrates the effects of severe drought conditions during 1985-86 (Bergman, et al. 1986), causing prolonged low flows in 1986. After above-average precipitation in late 1986, dry conditions returned in 1987, leading to flows 25-50% below median by mid-1987.

Thus, all areas of the Mississippi Basin experienced low flows during most of 1987. Indeed, consistently warm and dry conditions from January to June 1987 in the Midwest led to prohibition of barges from

## Table 4.1
## Classification of Average Temperature
## and Precipitation Conditions
## Prevailing in Three Regions During 1987-1988

| | Missouri Basin | | Upper Mississippi and Ohio Basins | | Tennessee and Lower Mississippi Basins | |
|---|---|---|---|---|---|---|
| | Temp.[1] | Precip.[2] | Temp. | Precip. | Temp. | Precip. |
| **1987** | | | | | | |
| Jan. | A | B | A | B | N | B |
| Feb. | A | N | A | B | A | N |
| Mar. | A | A | A | B | A | B |
| Apr. | A | B | A | B | N | B |
| May | A | N | A | B | A | B |
| June | A | B | A | B | A | N |
| July | A | N | A | N | N | N |
| Aug. | B | A | N | A | A | B |
| Sept. | N | B | N | B | N | B |
| Oct. | B | B | B | B | B | B |
| Nov. | A | B | A | A | A | N |
| Dec. | A | A | A | A | A | A |
| SUMS | | | | | | |
| A= | 9 | 3 | 9 | 3 | 7 | 1 |
| N= | 1 | 3 | 2 | 1 | 4 | 4 |
| B= | 2 | 6 | 1 | 8 | 1 | 7 |
| **1988** | | | | | | |
| Jan. | N | A | N | N | B | B |
| Feb. | N | B | B | B | N | B |
| Mar. | A | B | A | B | N | B |
| Apr. | A | B | N | B | N | B |
| May | A | B | A | B | N | B |
| June | A | B | A | B | N | N |
| July | A | B | A | B | N | N |
| Aug. | A | B | A | N | A | B |
| SUMS | | | | | | |
| A= | 6 | 1 | 5 | 0 | 1 | 0 |
| N= | 2 | 0 | 2 | 2 | 6 | 1 |
| B= | 0 | 7 | 1 | 6 | 1 | 7 |

1.  A = >2°F above normal; B = >2°F below normal; N = ±2°F of normal.
2.  A = >125% of normal; B = <75% of normal; N = ±25% of normal.

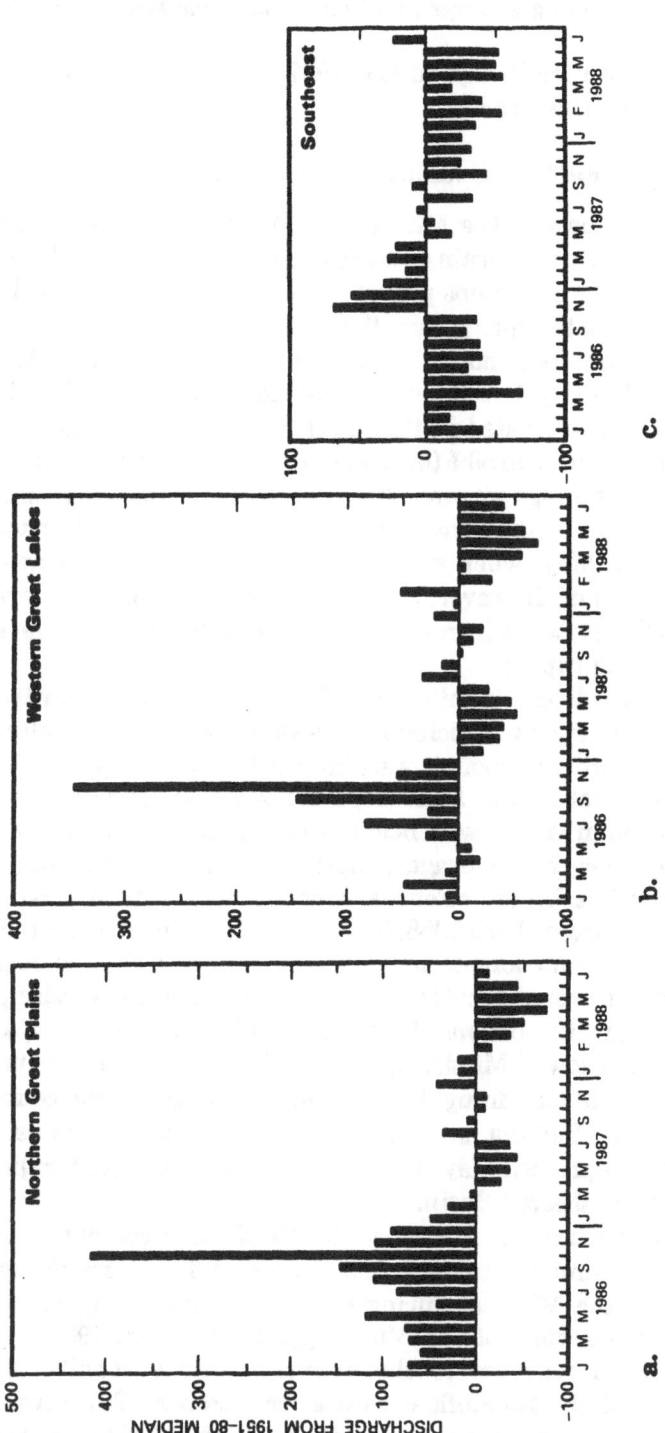

**Figure 4.2**

Area-mean streamflow in three major regions comprising most of the Mississippi River Basin, expressed as a percent of the 1951-80 median (*Source:* Midwestern Climate Center)

the Mississippi River for 10 days in July 1987 when once-in-10-year low flows occurred (Koellner 1988).

## Principal Factors Causing the Low Flows in 1988

The low 1988 flows in the Mississippi, Missouri, and Ohio River systems, which collectively drain 40% of the United States, were largely produced by weather conditions prevalent in early 1988. Snowmelt is a key contribution to the spring river flows of the upper Mississippi and Ohio rivers, but this was not to be the case in the spring of 1988. Snowfalls after January 1, 1988, were light and infrequent. The 1988 snowfall in the states located in the upper Mississippi drainage area ranged from 57% (Illinois) to 89% (Minnesota) of their long-term averages, and the basin-wide average was only 70% of normal. The highest monthly median streamflows in the central U.S. typically occur in February, March, April, and May, when snowmelt combines with the heaviest monthly precipitation. However, as shown in Figure 4.3, river levels in the western Great Lakes increased very little in March, when levels are normally rising rapidly.

The other major determinant of the 1988 low flows was a relatively warm and very dry spring. Precipitation was below normal in all the major basins in February; amounts were 25-75% below average over 90% of the northern Great Plains, Midwest, and Southeast. The dry, warm weather continued in March, with below-normal precipitation over 68% of the basin. Higher than average temperatures in the upper Mississippi and Missouri basins after February increased evapotranspiration. Thus, at the beginning of April 1988, the Palmer Drought Severity Index (PDSI), which represents long-term moisture conditions, showed either moderate or severe drought in: (1) the upper portions of the Mississippi and Missouri Rivers, (2) portions of the Ohio and Tennessee Rivers, and (3) portions of the lower Mississippi and Arkansas Rivers. By mid-May, areas of very severe drought had expanded to include the central area: Iowa, Illinois, Indiana, and Ohio. Precipitation was much below normal (>50%) in April and May, and dryness extended across all regions comprising the Mississippi Basin.

The drought continued to worsen, with 83% of the Mississippi Basin covered by severe drought by mid-June (Figure 4.3). Severe drought seldom exists in the Midwest during spring and, more importantly, rarely does it cover most of the Mississippi Basin (Karl 1988). The Mississippi River's response to the extreme climate conditions is revealed by the daily streamflow values for October 1987 through September 1988, as measured at two locations. Flows at Keokuk, Iowa (in the middle of the upper Mississippi River Basin), and at Vicksburg,

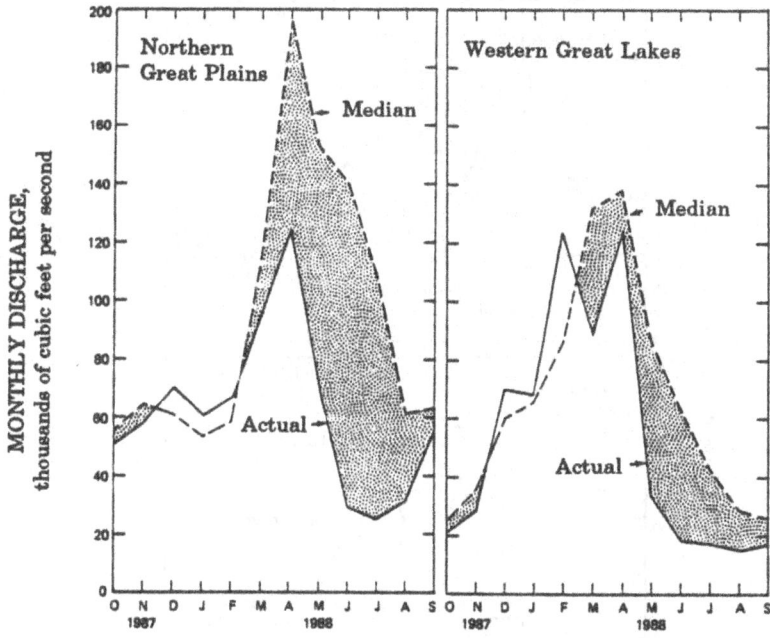

**Figure 4.3**
Monthly mean streamflows for October 1987-September 1988 and median flows (1951-80 base period) for two major regions (*Source:* Midwestern Climate Center)

Mississippi (the middle of the lower Mississippi River Basin), are depicted in Figure 4.4, along with their record extreme monthly average discharges. Deficient snowmelt from February to March caused flows to decline in late March. Heavy early April rains in parts of the basin are reflected in abrupt flow increases during early April (Figure 4.4). Thereafter, deficient rainfall brought rapid declines in flow. Normally, this is a time when flows increase toward normal annual peaks in April and May. Thus, the rapid decline in April was a strong indication that there would be very low flows in coming months. Before the end of May, the flows at both locations had reached record-low levels that were to continue throughout the summer.

Another factor in the very low flows is the lack of river control along the lower few miles of the Ohio River and the lower Mississippi River; all barge depth problems occurred where there are no lock and dam systems. Most of the Ohio, Missouri, and Upper Mississippi rivers have locks and dams operated by the Army Corps of Engineers to con-

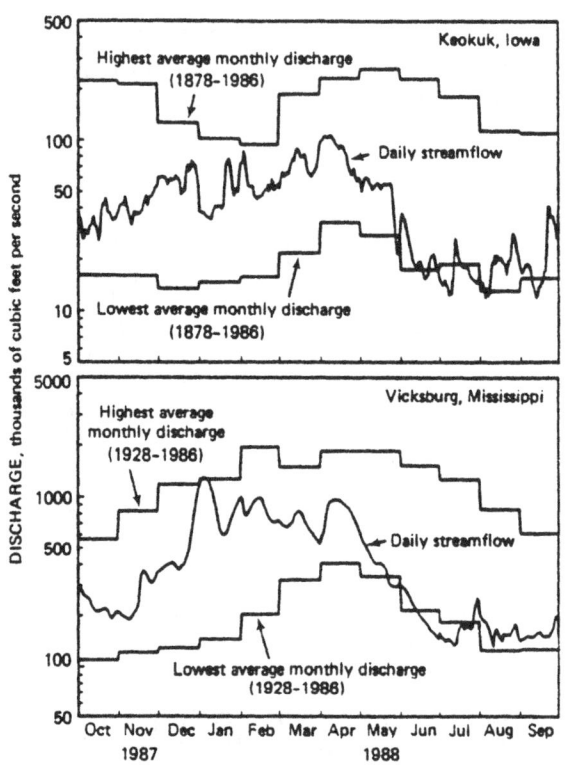

**Figure 4.4**
Daily streamflows on the upper (Keokuk) and lower (Vicksburg)
Mississippi River from October 1987 to September 1988 (1988 water year)
(*Source*: Midwestern Climate Center)

trol flooding and sustain sufficient water in the pools to maintain
adequate (≥9 feet, 2.7 m) river levels for barge transportation.
Additionally, these systems are used to control flows for transportation,
water supplies, hydroelectric power generation, and other uses
(Koellner 1988). Army Corps of Engineers releases of water in these
systems during 1988 kept river levels in the lower reaches of both
rivers, from just north of Cairo southwards, from falling lower than
they did (see Figure 4.3). Dredging is used routinely to maintain
channels for navigation in these uncontrolled areas of the lower
Mississippi and Ohio.

## Impacts of the Drought

The drought caused rapidly decreasing flows in the lower portions of the Ohio and the lower half of the Mississippi River by the end of May. In several channels near and south of Cairo, Illinois, the river depth fell to less than eight feet (2.4 m) by early June. Barges ran aground because the slowly moving rivers began depositing sediment in some channels, or "shoaling," and thus making the rivers more shallow.

### River Transportation Impacts

The first tow grounded on June 8, 1988, on the Mississippi River south of St. Louis. The Corps of Engineers dredged the area, and the Coast Guard limited the area to vessels drafting no more than six feet (1.8 m) (USACE 1988). By June 15, the level of the river passing Memphis was the lowest since records began in 1872. A 10-mile (16-km) stretch of the Ohio River from Cairo to Mound City, Illinois (just north of the confluence with the Mississippi), also experienced shoaling and river levels less than eight feet (2.4 m) by June 14 (*News-Gazette* June 20).

Fully loaded barges must have nine feet (2.7 m) of water in order to move. Thus, by mid-June, large numbers of tows and barges were halted in the Mound City area of the Ohio River, and near Greenville, Mississippi, and Memphis, Tennessee, along the lower Mississippi River (Figure 4.3).

On June 14, 1988, the Coast Guard closed a stretch of the Ohio River north of Cairo. More than 700 barges were backed up at Mound City, at a time when the daily cost for a tow pulling 20 barges ranged from $5,000 to $10,000 (Interagency Drought Policy Committee July 15, 1988). Intensive dredging began on June 14, and the river was reopened by June 17.

Mound City is an important river port where three firms load midwestern grain on barges. Inability to maintain barge movements and subsequent lack of empty barges forced the port to find storage for 200,000 bushels of grain. By June 27, more than $1 million worth of grain was held in open storage on Mound City streets, because elevators could not cope with the grain influx (*News-Gazette* June 27).[1] This situation confirmed Koellner's earlier prediction that low Mississippi system water levels could cause many tows and barges to be stranded near areas used for fleeting and loading (1988), and that port storage systems would prove inadequate to handle incoming shipments.

By June 17, 700 barges were backed up at Greenville, Mississippi, and dredging had begun there to clear a 2,000-foot (600-m) channel. By

June 19, 130 tows and 3,900 barges were backed up in the Mississippi River at Memphis, but dredging temporarily opened the blockage on June 20. Barge traffic was again halted in the Cairo area of the Ohio River on June 27, and 2,000 barges were held up by low flows for several days in early July at Memphis (*Farm Week* July 4). Other blockages occurred along the lower Mississippi at seven locations south of Cairo (Helpa 1988).

*Economic Impacts and Responses*

Blockages in the Ohio and Mississippi rivers created problems for bulk commodities shippers and the commodities market. By early July, river traffic was down 20%, and loads totaling 30 million tons (27.3 million metric tons) were halted (Helpa 1988). Barge and tow owners and shippers experienced immediate economic losses, while most river ports along the central and upper river system had to deal with reduced shipments and commodity backups, such as those at Mound City.

As the difficulties of moving loads became widely known in mid-June, transport of many commodities was shifted to railroads. Other loads were moved north to Great Lakes ports, creating a serious loss of business for the river ports. By late July, the river flows had increased sufficiently from heavy July rainfalls in the eastern Corn Belt to prevent further major blockages (Figure 4.4). However, the flow of the Mississippi at Vicksburg on August 11, 1988 was only 80,000 cfs (3,920 m$^3$/sec), as compared to a normal of 320,000 cfs (8,960 m$^3$/sec), and barge loads remained less than average (*Chicago Tribune* August 2).

The low flows produced other notable impacts, including a 25% decrease in hydropower generation, a 15% decrease in recreational uses of rivers and lakes, and salt water intrusion 105 miles (168 km) up the Mississippi River past New Orleans (Helpa 1988).

## Management Responses

As might be expected from the severity of the river transport problems, several major responses quickly emerged. In June, the River Industry Executive Task Force was formed, and was comprised of representatives from the Army Corps of Engineers and the Coast Guard, and leaders of private companies (shippers, tow owners, and port managers). The American Waterways Operators convened the task force in Washington, D.C., in late June to jointly identify the best responses to the low flows (American Waterways Operators October 28, 1988).

The impacts of the drought and responses by both the private sector and the agencies represented in the task force are best examined in a historical context. Equally low flows in the lower half of the Mississippi River occurred in the 1930s and mid-1950s, but the tow and barge industry was small in the 1930s and just beginning its marked growth in the 1950s (Koellner 1988). The 9-foot (2.7-m) channelization of the Mississippi River (south of Minneapolis) was not completed until 1939. In 1950, the Mississippi system carried only 8 million tons (7.2 metric tons) of commodities; by 1980 this had increased to 100 million tons (90.9 metric tons). Hence, the 1988 low flows were the first major climatic stress this industry had encountered since becoming a significant part of the bulk commodity transportation network of the central U.S., and the responses had to be assessed in this light.

The event was not totally unanticipated, however. A Corps of Engineers staff member, asked in 1987 to consider possible problems related to a climate change leading to drier conditions in the Midwest, had described essentially all of the problems of 1988 (Koellner 1988). But no drought contingency plans for the lower Mississippi River were in place, leaving the task force response to be formed under crisis conditions.

The first action was to *dredge* blocked river reaches. The Army Corps of Engineers dredged 11 blocked areas, but efforts were concentrated on three primary trouble areas—one near Cairo on the Ohio River, and two along the Mississippi River. Dredging activities at any one blockage typically involved several days of constant work by several dredges. For example, four days of round-the-clock dredging were needed to create a 2,100-foot channel (630 m), 300 feet wide (90 m) and 11 feet deep (3.3 m), near Cairo between June 14 and 17, and 13 dredges operated continuously during the last two weeks of June and throughout July to maintain navigable channels along the Mississippi (Helpa 1988).

The second response was to *reduce the number of tows and the size of barge loads.* On June 23, 1988, the Coast Guard ordered reduced numbers of barges per tow on the Mississippi and Ohio rivers. Thirty to 40 barges per tow are a typical load south of Cairo, but the order set the maximum number at 20 per tow. Subsequently, the Coast Guard issued tonnage restrictions for barges (*News-Gazette* July 9). The result was fewer barge movements with lesser loads and greatly reduced bulk commodities shipping throughout the system, adversely affecting barge and tow companies, river ports, shippers, and producers (cf. Chapter 3). The Coast Guard lifted its tow-size restrictions in September, but issued advisories recommending that reduced tow sizes and loads continue during the fall of 1988.

A third response was the *emergency management of water releases* from lock and dam systems along the Missouri, upper Mississippi, and

Ohio rivers. The Corps of Engineers carefully balanced water releases from each in order to release as much water as possible to the lower reaches, but also to meet the water requirements in the controlled reaches of the rivers.

Fourth, more shipments were diverted to the *Tennessee-Tombigbee Waterway*, a system built and operated by the Corps of Engineers. It parallels the Mississippi from Cairo to New Orleans, but is not as favored as the Mississippi, because the speed and direction of the river currents do not aid southward movement of loaded barges as much as the flow of the Mississippi. Traffic was diverted to this waterway in June, and barge cargoes on the waterway increased to 2.1 million tons (1.82 million metric tons) in July 1988, compared to 300,000 tons (272,000 metric tons) in July 1987 (Helpa 1988).

The negative impacts on the barge and tow companies created an *increase in barge shipping rates*. During June 9-16, barge shipping rates were raised from $5 a ton (909 kg) for bulk commodities to $14 or $15 per ton (909 kg), due to reduced tonnage on barges and longer travel times (*Farm Week* June 20). This increase made shipping by other means (Great Lakes ships and railroads) unusually competitive with the barges.

There were also increases in the use of *alternative means of transportation.* The most immediate were railroads that served the same commodity-producing areas and connected local producers to ports along the major rivers. Consequently, the Great Lakes ports and shipping industries became means of moving exportable commodities, especially grain, out of the Midwest, because railroads moved the grain to the ports of the Great Lakes, instead of New Orleans.

The Illinois Central Railroad (ICRR), which is a north-south system connecting the Great Lakes at Chicago with New Orleans (and thus parallels the heavily used Illinois River-Mississippi River barge system and its major ports), played a key role in responding to the river blockages. The ICRR and the tow-barge industry are major competitors, and the railroad continually monitors barge rates, as well as present and future river conditions. In March 1988, a Corps of Engineers spokesperson reported that river levels could drop to a crisis level if spring precipitation did not dramatically increase. As a result, ICRR employed a private weather and climate forecasting firm to provide them with an outlook for spring precipitation (*Chicago Sun Times* July 24). Their analysts predicted the worsening spring and summer drought in April, and the railroad's management team concluded that there would be very low river flows, severely limiting barge shipments.[2]

Believing the prediction, the railroad decided in early May to begin stockpiling large (100-ton, 90.9 metric tons) hopper cars to haul extra coal and grain. By June they had leased more than 521 hopper cars, at a cost of $700,000, to supplement the railroad's fleet of 4,000 (*Chicago Tribune* August 2). The railroads charged $8-$12 per ton (909 kg) to carry grain in 1988, and thus gained a competitive advantage over the barge operators, who had increased prices to $14-$15 per ton because of the low flows. The Illinois Central Railroad normally loses money in coal and grains during low summer traffic, but the increased business during the summer of 1988 led to its first profit for July since 1851! The ICRR hauled 9,201 grain cars to New Orleans in July, or 4,801 more than in July 1987, and carried 12,972 cars of coal—4,361 more than July 1987 (*News-Gazette* August 3).

This profitable experience caused the railroad to reassess its competitive status with barges and to consider future challenges to the barge industry. The ICRR climate advisors issued long-range outlooks for continued dry conditions in 1989, with low flows during the winter of 1988-89, the potential for more easily frozen shallow rivers, and continued reduced barge traffic (Peterson 1988). Thus the railroad retained the leased hopper cars through the spring of 1989 (*News-Gazette* August 3).

Great Lakes shippers and midwestern ports also took advantage of the barge problems. They, too, gained economic competitiveness from the unreliable movement of river barges and higher shipping costs for moving grain to New Orleans. The diversion of shipments to the Great Lakes was sizeable enough for a spokesperson from the Illinois International Port at Chicago to call the drought "a windfall" (*Chicago Sun Times* July 24). By late July, the port had shipped $1.87 million worth of grain that would have normally gone down the Mississippi River to New Orleans, gaining an addded income of $0.5 million. By mid-August, the shipping on the St. Lawrence Seaway had increased 7% above average due to diverted river traffic shipments.

## A Controversial Response Proposal: Increased Water Diversion

The problems on midwestern rivers led to what was perhaps the most controversial proposed drought response in 1988. As the low flows and barge problems became severe in early June, the Metropolitan Sanitary District of Greater Chicago (MSDGC) suggested increased water diversion from Lake Michigan at Chicago. The increased flow would move down the Illinois River to enter the Mississippi River near St. Louis (*Windsor Star* July 19). The plan was to increase the current diversion, which is limited to 3,200 cfs (89.6 m$^3$/sec) by U.S. Supreme Court decree, to 10,000 cfs (280 m$^3$/sec) for a 100-day period after obtaining an emergency declaration by President Reagan or the U.S. Supreme Court (Kudrna, et al. 1980).

This proposal, though controversial, was not without precedent; added diversion to increase Mississippi River flows was implemented during the severe 1953-56 drought. At that time, the MSDGC first allowed an increased diversion (up to the U.S. Supreme Court's limit) for 10 days in October 1956. In December, the Supreme Court approved a request from Illinois for an emergency increase in the diversion of 10,000 cfs for 100 days (the same as their request in 1988), but modified it to 8,500 cfs (238 m³/sec) for 76 days.

*Political Feasibility*

At a meeting of state officials on June 22, 1988, the Illinois Director of Agriculture reportedly proposed that the Army Corps of Engineers be directed to channel additional water to aid barge traffic through the Illinois River to the Mississippi (*News-Gazette* June 23). On June 23, Illinois Governor James Thompson proposed the plan to triple the diversion at the National Governor's Association Drought Conference in Chicago (*Chicago Tribune* June 24). Thompson's plan was to have the Corps of Engineers increase the diversion to 10,000 cfs (280 m³/sec), which was expected to raise the Mississippi River level at St. Louis by 1 foot (30 cm) and at Memphis by 6 inches (15 cm). It was also expected that this action would only lower Lake Michigan by about one inch (2.5 cm).

Illinois officials were confident that the plan was technically feasible, but they badly over-estimated its political viability. There is a long history of conflicts between Illinois and the other lake states (and Canada indirectly) over the amount of lake water diversion at Chicago (Kudrna, et al. 1980). Although the current allowable diversion is fixed by long-standing U.S. Supreme Court decisions, proponents of the 1988 proposal, including the states along the lower Mississippi River, apparently hoped that an increase as an "emergency measure" would gain support from Great Lakes states. However, when the plan was proposed at the National Governor's Conference, most other lake state governors immediately and strenuously objected (*Chicago Tribune* June 24), and operators of major ports around Lake Michigan expressed anger and outrage (*News-Gazette* July 9). *Farm Week* (July 18), an influential agricultural newspaper, simply called the plan a "political hot potato" that was unlikely to gain sufficient support.

The director of the Seaway Port Authority in Duluth, which ships wheat from the Great Plains, was one of many who spoke strongly against the plan. He felt that "every inch of water in the Great Lakes is essential to navigation and reacting to the proposal favorably would create a litigation nightmare" (*News-Gazette* July 9).

Those supporting the diversion included the Metropolitan Sanitary District of Greater Chicago, the Governor of Illinois, certain members of Congress, barge firms, certain grain and coal shippers, and the American Waterways Operators (*News-Gazette* July 9). However, the American Waterways Operators shifted to an antidiversion position in late July (*News-Gazette* August 1), stating that July conditions were not serious enough to warrant diversion.

Diversion proponents appeared to ignore a critical factor that fueled objections. At the time of the proposed diversion, the level of Lake Superior was eight inches (3.1 cm) below its long-term average, having fallen 1.3 feet (0.39 m) from record high levels in less than two years. The other Great Lakes had also fallen dramatically. Lakes Michigan and Huron (from whence the proposed increased diversion would occur) had fallen 2.6 feet (0.78 m) from record high levels that had persisted until January 1987 (Changnon 1987). Studies based on historical climate scenarios had indicated that Lake Michigan would take four years or longer to fall from its record heights of 1986 to average levels (Hartmann 1988), but in less than 18 months (January 1987-June 1988) the lake fell below its average level because of the 1987-88 drought! Illinois decision makers either did not know of this information or seriously underestimated its impact on public opinion regarding diversion.

In the ensuing political controversy, governors of four states (Wisconsin, Michigan, Indiana, and Ohio) threatened court action, and Illinois citizens living along the Illinois River objected, fearing that the tripled diversion would flood valuable lowlands (*News-Gazette* July 9). Yet, on July 8, 13 senators from Illinois and several southern states formally asked President Reagan to authorize the emergency diversion (*News-Gazette* July 9). Other senators from Ohio and New York strongly opposed the proposal. The Corps of Engineers was asked to quickly finish its study of the Illinois plan, begun on June 24, in order to decide whether to begin the complex process of seeking permission for the diversion from the U.S. Supreme Court.

During this deliberation, Canada's Environment Minister McMillan called the Illinois plan an "insane idea" (*News-Gazette* July 9), and several Canadian newspapers carried articles expressing outrage. The Canadian Ambassador to the United States delivered a formal note to the U.S. State Department on July 9, 1988, stating that "Canada was unalterably opposed to the diversion." The diversion plan and its opposition had thus become an international controversy. Illinois officials admitted that such strong opposition hurt the chances of winning approval for the diversion (*News-Gazette* July 9).

## Diversion Denied

On July 14, the Secretary of the Army for Civil Works declined the Illinois request, indicating that "there was no reason now or in the foreseeable future to increase the amount of water diverted out of Lake Michigan to enhance navigation on the Mississippi River" (Interagency Drought Policy Committee August 12). The USACE study had concluded that there would be little improvement in river channel conditions with a diversion of 10,000 cfs. The news media reported that the decision was based on an engineering analysis concluding that the added water was not needed, and that the decision was not politically motivated (*Farm Week* July 18).

Opposition from the Canadians, other lake states, and Illinois citizens was foreseeable. Why Illinois offered the plan in the face of such likely and strong opposition is a matter to be questioned. One possibility is that diversion supporters lacked knowledge about the widespread perception that the water-supply problem was serious in the Great Lakes, and failed to take into account the rapid decline in lake levels caused by the drought. This situation solidified the negative attitudes of other lake states and Canada.

Another possibility is that Illinois based its decision to pursue the diversion on an engineering and historical perspective, and felt the plan would be acceptable because it would have had only a slight effect on the net basin supply and the level of Lake Michigan. From this standpoint, it is understandable why they expected only minor objections. Furthermore, the success of a similar Illinois proposal in 1956 suggested acceptance for the proposal in 1988. However, the rapidly declining lake levels in June 1988 and the new water protectionist theme set by the Great Lakes Charter (1985) argued against its acceptance.

Regardless of the reasons for rejection, the diversion proposal may serve as a harbinger of the type of political controversies that will result from any type of proposed diversion increase for any location in the basin, particularly given the drier future climate conditions expected in the central U.S. (Koellner 1988).

## Implications for the Future

The complex and inter-related set of drought influences and responses in the Midwest, especially the controversial proposed diversion, offers several lessons about droughts and their management. First, the use by the ICRR of a climate prediction in a major economic decision (with a very positive outcome) favorably impressed management with the value of climatologically based decision making. ICRR leaders recognize they took a risk, and their decision to retain leased hopper cars through the 1988-89 winter based on other "climate trend" outlooks may not prove as profitable. Conversely, neither the Army Corps of Engineers nor barge companies anticipated the intensification of the drought and the concomitant river-flow problems. Yet, in hindsight it is obvious that a strategic plan, coupled with greater attention to indicators of impending low flows that were evident by the rapidly falling levels in April, could have reduced the impacts through earlier reactions (i.e., added dredging) and more cost-effective responses such as earlier diversion of traffic to the Tennessee-Tombigbee Waterway.

Second, it appears that hydroclimatic information about the severity of the drought and, in particular, the serious decreases in Great Lakes water levels, was ignored in the decision to propose the diversion plan. The proposal may have been hydrologically and technically feasible, but was politically flawed. Several political issues were overlooked, such as the fact that the diversion amount had been contested for 60 years (Kudrna, et al. 1980). Opponents feared that even an "emergency" diversion would have implications beyond the drought. In addition, the proposal came only three years after all the governors of the lake states had agreed to protect Great Lakes water supplies in the Great Lakes Charter (1985). Of further interest is the fact that Illinois had not proposed an increase in the diversion during the record high lake levels of 1985-86 (Changnon 1987) to help reduce shoreline damages.

The 1988 controversy also reflects societal sensitivity to the water resources of the Great Lakes and elsewhere throughout the humid eastern U.S. (Changnon 1987). It illustrates the likelihood of future controversies if the climate becomes drier, as suggested by some $CO_2$ scenarios. Policy makers and scientists concerned about the Greenhouse Effect should take note of the controversy.

The drought also revealed a lack of planning by industry and relevant public sector entities. Crisis management, rather than strategic planning, was widespread. Drought contingency planning by the Army Corps of Engineers in the past had been on a project-by-project basis. When the 1988 drought developed, the Corps of Engineers was developing more integrated drought plans for the southeastern U.S. and selected river basins, but there was no basin-wide plan for the Mississippi River. Helpa (1988) reported that in the fall of 1988, the Corps was anticipating future low flows (and related added dredging) during the winter of 1988-89, and had asked Congress for funding to conduct additional studies to develop drought contingency plans for all areas under its control.

While these are encouraging developments, many losses could have been averted in 1988 had contingency plans existed, along with better use of the hydroclimatological data that were available.

Another area of adjustments concerns possible shifts in shipping patterns and costs. The Illinois Central Railroad reacted quickly and effectively to the economic advantages occasioned by the drought, and began planning for permanent future changes in shipping (*News-Gazette* August 3). Based on the outlooks of two private forecasters, the climate outlooks of other scientists, and widely circulated statements in 1988 that the greenhouse effect will soon lead to warmer and drier conditions, ICRR advisors predicted that dry conditions (and low river flows) could be expected for another five to 10 years (Peterson 1988).

Expectations of drier conditions have caused regional railroads and Great Lakes shippers to reassess their rates for bulk commodities in order to gain a greater competitive advantage in attracting shippers of grain, coal, and other bulk commodities (*News-Gazette* August 3). However, barge industry leaders anticipated no permanent loss of business from the 1988 drought, even though October barge rates were still 8-10% higher than in 1987 (President's Interagency Drought Policy Committee 1988). A campaign to compete more widely for bulk commodity shipments in the Midwest apparently is emerging as a result of the 1988 drought.

The 1988 events in the Midwest also showed the value of rapid integration of weather information that signals significant changes to the physical environment, as well as the need for early and continuing issuance of drought status reports. The 1987-89 drought illustrates the value that can be derived by use of probabilistic seasonal predictions. But, it also shows how various sources of climate prediction can be combined to derive a singular long-term outlook, such as that created for the ICRR, that would be considered highly questionable by most of the scientific community. Regardless, the great difference in the reactions to the waterways problems and the planned responses of the Illinois Central Railroad underscores the value of climate monitoring and the wise use of long-range climate predictions.

### Notes

1. One barge hauls the equivalent of the load contained in 15 railroad cars or 60 semi-trailer trucks (American Waterways Operators 1988).

2. Based on correspondence from G.F. Mohan, Senior Vice President, Illinois Central Railroad.

## References

American Waterways Operators. *Barge Industry Reveals Drought Economic Impacts*. Arlington. August 1, 1988.

——. "RIETF Meets October 18." *AWO Letter* 45, 22: 3. 1988.

Bergman, K.H., C.F. Ropelewski, and M.S. Halpert. "The Record Southeast Drought of 1986," *Weatherwise* 39: 262-266. 1986.

*Champaign-Urbana News-Gazette*, June 17, 1988: "Dry Spell Hurts River Traffic."

——, June 20, 1988: "Barges Resume Trips."

——, June 23, 1988: "Barges Halt."

——, June 27, 1988: "Delays in Barge Traffic Cause Million Dollar Grain Pileup."

——, July 9, 1988: "Great Lakes Diversion Proposal Finding Little Government Backing."

——, August 1, 1988: "Barge Operators Estimate Losses From Drought."

——, August 3, 1988: "Railroad's Profits Rolling in During Summer Drought."

Changnon, S.A. "Hydrologic Applications of Weather and Climate Information." *Journal of the American Water Works Association* 10: 514-518. 1981.

——, "Climate Fluctuations and Record High Levels of Lake Michigan." *Bulletin of the American Meteorological Society* 68: 1394-1402. 1987.

Changnon, S.A., S. Sonka, and S. Hofing. "Assessing Climate Information Use in Agribusiness. Part 1: Actual and Potential Use and Impediments to Usage." *Journal of Climate* 1: 757-765. 1988.

*Chicago Sun Times*, June 26, 1988: "Lake Level Heading Towards a 12-Year Low."

——, July 24, 1988: "Drought's Long Shadow."

——, June 24, 1988: "Four Governors Oppose Thompson Water Plan."

——, June 24, 1988: "Greenhouse Effect Has Arrived, Expert Says."

——, August 2, 1988: "Drought Toll, Opportunity."

*Farm Week*, June 20, 1988: "River Traffic Latest Victim of Drought in the Breadbasket."

——, July 4, 1988: "Channel Shipping Blocked in Rivers."

——, July 18, 1988: "No to Lake Diversion Plan."

Great Lakes Charter. *Principals for Management of the Great Lakes Water Resources*. Madison: Council of Great Lakes Governors. 1985.

Hartmann, H. *Potential Variation in Great Lakes Water Levels: A Hydrologic Response Analysis*. NOAA Technical Memorandum ERL, GLERL 68. Ann Arbor. 1988.

Heim, R.R. "About That Drought." *Weatherwise* 32: 266-271. 1988.

Helpa, M.J. "Corps of Engineers Briefing on Drought." In *The Drought of 1988 and Beyond*. Rockville, Maryland: National Climate Program Office. 1988.

Interagency Drought Policy Committee. *Report on Drought Conditions*. Washington, D.C.: White House. July 15, 1988.

——, *Report on Drought Conditions*. Washington, D.C.: White House. August 12, 1988.

——, *The Drought of 1988*. Washington, D.C.: White House. December 1988.

Karl, T. "Temporal and Spatial Severity of the 1988 Drought: A Historical Perspective." Paper presented at the Conference on Strategic Planning for Droughts, Washington, D.C. 1988.

Koellner, W. "Climate Variability and the Mississippi River." In: *Societal Responses to Regional Climatic Change, Forecasting by Analogy.* Edited by M.H. Glantz. Boulder: Westview Press. 1988.

Kudrna, F.L., D. Vonnahme, and K.L. Brewster. "Lake Michigan Water Allocation." *Journal of Water Resources Planning and Management* 106: WRI:43. 1980.

Peterson, J.B. and Associates. *An Appraisal of Climatic Trends Affecting Mississippi River Levels and Long-Term Outlook.* Chicago. 1988.

*Time,* July 4, 1988: "The Big Dry."

U.S. Army Corps of Engineers. *Situation Report,* DAEN-CWO-M. Washington, D.C.: Office of Chief of Engineers. June 9, 1988.

*Windsor Star,* July 19, 1988: "Thirsty South Keeps Trying to Turn On Great Lakes Tap."

# 5

# Drought and Dryland Agriculture in North Dakota

The northern U.S. Great Plains—where the majority of the country's spring and durum wheat is grown—and adjacent parts of Canada were seriously affected by the 1987-89 drought. North Dakota dryland farmers were especially hard hit. Several factors make this event of special interest. It was the second severe drought to affect the area in the 1980s, and it seriously reduced both 1988 and 1989 wheat yields. Yet, North Dakota farm income held steady in both years, suggesting the sometimes counter-intuitive effects of drought within the modern economic and policy context of Great Plains dryland agriculture. The drought also occurred at a time when public concern had been raised over the potential for global climate change through the Greenhouse Effect. If farmers link the 1987-89 drought to climate change, then it may have greater influence on their strategies over the long-term than previous droughts. In examining the drought in North Dakota, we first describe its physical and social impacts on the region and then explore farmers' drought perceptions, responses, and their views of the drought vis-à-vis climate change. Finally, we discuss the effect of the drought on agricultural stability in the broader context of Great Plains farming.

## The North Dakota Drought

Severe drought marked both the 1988 and 1989 growing seasons in North Dakota—heart of the country's spring wheat production area. The drought was the second severe dry spell in the 1980s. It was the fourth serious drought in the last three decades (dry spells occurred in 1961, 1978, and 1980). Although the region experienced fringe effects of the 1950s southern Plains drought, only the well-known Dust Bowl droughts in the mid-1930s were as severe as 1988 and 1989 in this region.

*Climatic Conditions*

The 1987-89 North Dakota drought was, climatologically, the second worst in the state's history and perhaps the worst in terms of total agricultural losses (Aakre, Leholm, and Leistritz 1988). Statewide drought indices (PHDI) became negative in late 1987, as in many other parts of the country, and the drought worsened rapidly through mid-summer 1988 (Figure 5.1). The statewide PDHI in July was lower than any year since 1936. Spring and summer precipitation was 8.20 inches (208.28 mm), compared to a 1951-80 average of 13.78 inches (350.01 mm); it was the third driest growing season on record, after 1934 and 1936. Precipitation in the state by late-August ranged between 40% and 70% of the 30-year mean, with an average of 57% of the mean statewide. Dry soils initially were a minor boon as the growing season started, allowing earlier and easier spring planting. Lack of precipitation and high evapotranspiration rates, however, subsequently produced critically low soil moisture conditions (Table 5.1). As elsewhere in the U.S., pre-existing dryness set the stage for the severe summer drought that emerged with a suddenness that surprised many longtime observers.

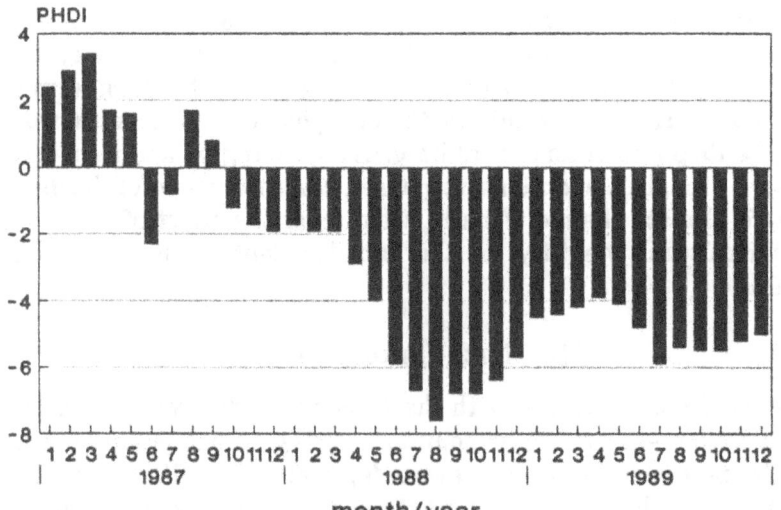

month/year

Figure 5.1

**North Dakota Statewide PHDI (Data from the North Dakota Agricultural Statistics Service)**

## Table 5.1
## Soil Moisture in North Dakota

| | % of North Dakota with: | | | |
| --- | --- | --- | --- | --- |
| | Very Short Moisture | Short Moisture | Adequate Moisture | Surplus Moisture |
| **Topsoil** | | | | |
| August 21, 1988 | 58 | 34 | 8 | 0 |
| August 14, 1989 | 60 | 33 | 7 | 0 |
| 1983-87 average | 14 | 35 | 47 | 4 |
| **Subsoil** | | | | |
| August 21, 1988 | 81 | 16 | 3 | 0 |
| August 14, 1989 | 74 | 21 | 5 | 0 |
| 1983-87 average | 11 | 35 | 54 | 0 |

*Source*: North Dakota Agricultural Statistics Service, *Crop Weather Bulletin*, August 22, 1988; Aug. 15, 1989.

The summer of 1989 was only slightly better, with hot, dry conditions redeveloping after a relatively dry winter (Figure 5.1). Growing season precipitation in 1989 was slightly better, at 9.07 inches or 66% of the 30-year mean, yet soil moisture by harvest-time was as poor as it was in 1988 (Table 5.1).

### Agricultural Impacts

North Dakota is well-suited in many ways to dryland wheat production. Some 60-70% of the annual precipitation typically falls during the growing season from April to September, and cool summer temperatures generally reduce the evapotranspiration stress so common in the remainder of the U.S. Great Plains (North Dakota State Water Commission 1987). Hard red spring wheat, the dominant crop, is valued for its milling qualities, and the durum wheats, grown almost exclusively in North Dakota, provide most of the country's pasta production. Spring and durum wheat yields in 1988 were 41% and 53% of average, respectively, and the all-wheat yield was the lowest since the extremely dry spring of 1961 (Figure 5.2; North Dakota Agricul-

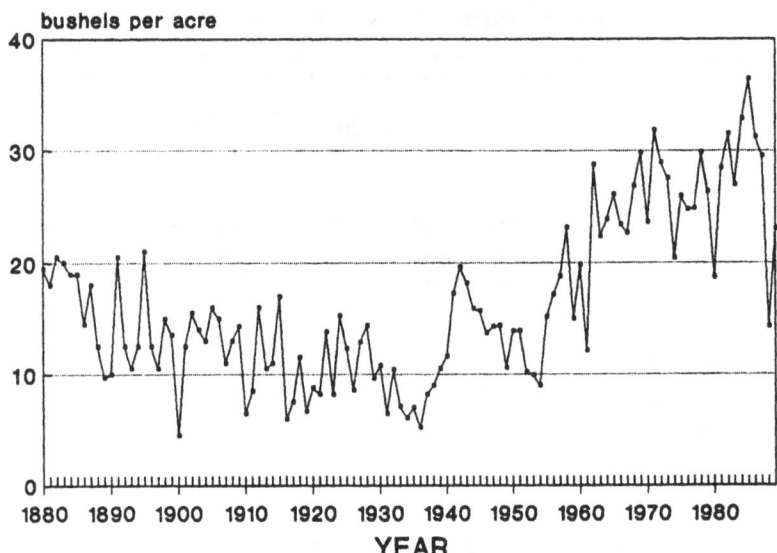

**Figure 5.2**
North Dakota Wheat Yields, 1880-1989 (Data from North Dakota
Agricultural Statistics Service)

tural Statistics Service 1988). The 1989 yield was not much better
(Figure 5.2).

While wheat yields have been steadily increasing with greater
fertilizer and other technological inputs since the 1930s, the absolute
and relative declines from the trend in recent drought years (i.e., 1961,
1980, and 1988) are larger than in the 1930s (Figure 5.2). This
suggests that, in some respects, modern dryland farming is as
physically susceptible to drought effects now as it was in the 1930s.

Statewide yield data obscure near-total wheat losses during 1988
in parts of central and western North Dakota (Figure 5.3), as well as
abandonment of a quarter of the planted fields in 1988. Thus, total
production was down more than 50% from previous years and lower
than the severe drought of 1980. Wheat yields were somewhat better
in 1989, and there was less abandonment, so production was almost
doubled (Figure 5.4).

The livestock industry in North Dakota was also hard hit. In mid-
August 1988, only 8% of the state's available hay and roughage
supplies were rated of adequate quality. Because 80% of the state's
pastures were rated poor to very poor, feed availability was down, and
feed prices were up, farmers were forced to sell their stock in large

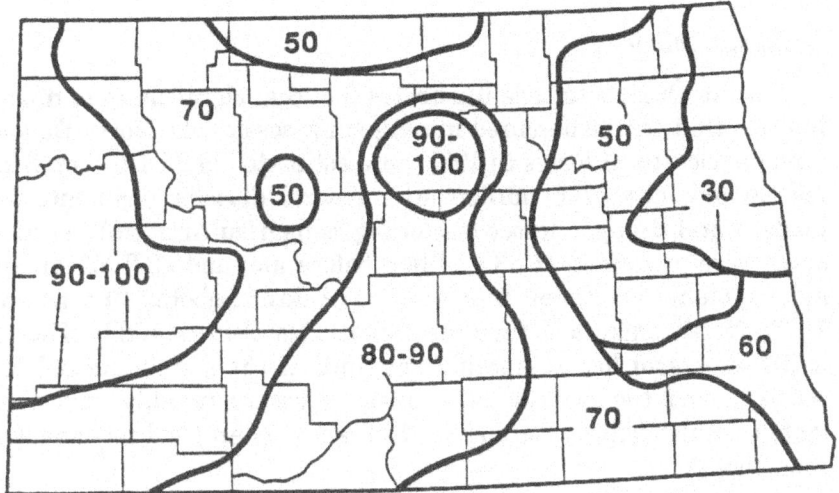

Figure 5.3
Percent of small grain crop loss in North Dakota by area as of June 30, 1988 (*Source*: Aakre, et al. 1988)

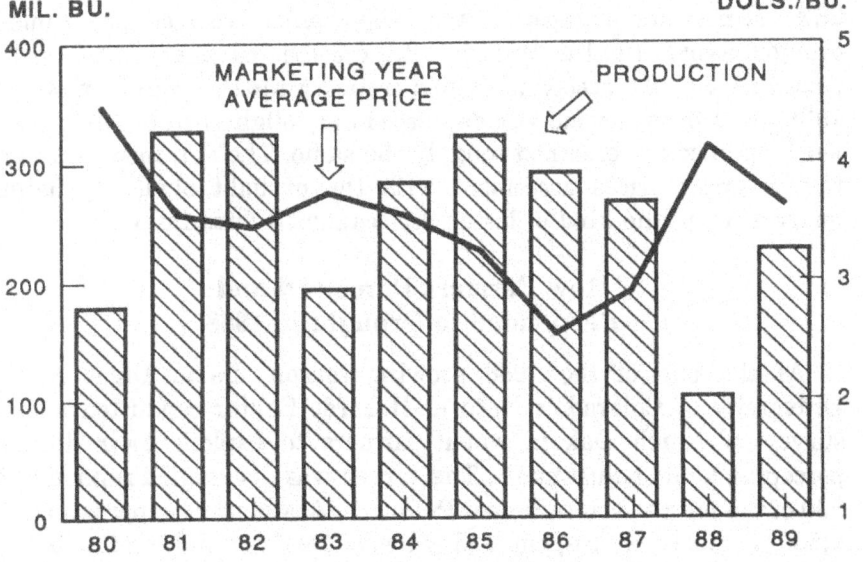

Figure 5.4
Production yields and prices for all wheats in North Dakota, 1980-89 (Data from North Dakota Agricultural Statistics Services)

numbers, setting a new record for cattle marketing (North Dakota Agricultural Statistics Service 1988, 1989).

*Economic Impacts*

The North Dakota case illustrates the local significance of drought impacts that can be obscured in national assessments. North Dakota's direct agricultural losses in 1988 were estimated at $1.1 billion before federal aid, and $706 million after federal disaster payments were made. Total drought losses (including nonagricultural effects) in the state were estimated at $3.4 billion before aid, and $2.2 billion after aid payments are taken into account (Aakre, Leholm, and Leistritz 1988; North Dakota Agricultural Statistics Service 1989). The pre-assistance farm loss represents one-third of the state's normal farm receipts, and the pre-aid total losses represent roughly 15% of the state's total business volume (or 10% after federal aid payments are considered)!

Despite these losses, farm income did not suffer. In 1989, Leistritz, et al. (1989) resurveyed 466 of the North Dakota farmers previously studied to ascertain how the drought had affected their operations and long-term stability. They found that grain crop losses averaged 71%, and that 45% of livestock producers reduced their herds by at least 25%. Ninety-one percent of the respondents received government drought assistance; the average total disaster payment was $15,234. About 61% of the respondents had crop insurance. Overall, however, 1988 gross farm income was depressed only slightly from 1987, while net farm income remained roughly the same. The aid, insurance, and higher grain prices associated with the drought obviously helped farmers avoid the kind of losses that can spell farm failure.

## How Farmers Perceived and Responded to Drought in 1988

At the end of the 1988 growing season, David Diggs at the University of Colorado's Natural Hazards Center conducted a mail survey of North Dakota wheat farmers to explore their drought perceptions and adjustments. The survey was distributed randomly to 300 of the North Dakota Wheat Producers Association's approximately 1,500 members; 92 responses (31%) were received and tabulated.

The respondents were generally older and experienced farmers who had an average of 26.6 years in farming. Over two-thirds were over the age of 40 and almost one-half were over 50 years of age. The farms they operated were relatively large, with a mean of 1,914 acres and a median of 1,600 (compared to a state mean of 1,252 acres). Respondents were from all parts of the state, but two-thirds were from the western half, where the largest wheat farms predominate.

## Drought Perception

Farmers viewed the 1988 drought as one of the worst in recent history and perhaps as a precursor of things to come. Respondents compared a large number of other drought years with 1988, but the ones most often mentioned included: 1961 (27 mentions), 1980 (26), 1981 (10), 1935 (8), and 1936 (6). In addition, only the 1961 drought was noted by many as being of the same severity as 1988.

Fifty-three percent of the respondents thought that droughts were becoming more frequent, 42% felt that droughts were occurring at about the same frequency, 5% did not know, and no respondent believed that droughts were becoming less frequent in the state. When asked what the chance of drought in 1989 was, 54.3% said greater than average, 37.0% said average, and 8.4% felt the chance was below average.

## Drought Adjustments

Farmers reported using a large and varied roster of drought adjustments in 1988 (Table 5.2), with an average of five different drought adjustments cited by each survey respondent. The roster included a mixture of both purely financial adjustments, such as accepting federal relief, reducing family expenses, and receiving insurance payments; and resource management adjustments, such as applying soil conservation techniques and using crop fields for forage.

When asked if the 1988 drought would affect their mid- to long-term (three or more years) farm management strategies, 39.3% said yes, 37.1% said no, and 23.6% did not know. A wide array of possible planning changes were cited, though most were financially oriented, such as delaying capital investments and replacements of machinery, planning more conservatively, and slowing expansion plans.

## Perceptions of the Drought and Climate Change

Farmers tended to link the 1988 drought to longer-term climate change. When asked if the farmers thought this drought was part of a change in the climate of the region, 40.4% (36 out of 89 respondents) said yes, 31.5% (28) said that they didn't know, and 28.1% (25) said no. When those who answered yes were asked to describe how the climate was changing, they almost unanimously checked *warmer*, *dryer*, and with slightly less frequency *longer growing season*. Asked why the

## Table 5.2
## Drought Adjustments Used by Farmers During 1988 Drought

| Adjustment | Freq. | % Citing Adjustment |
|---|---|---|
| Signed up for federal drought relief | 61 | 66.3 |
| Used soil conservation measures to reduce erosion | 60 | 65.2 |
| Reduced family expenses | 54 | 58.7 |
| Signed up for or received insurance payments | 52 | 56.5 |
| Used crop fields for forage | 40 | 43.5 |
| Tilled ground to conserve moisture | 37 | 40.2 |
| Sold stored grain | 35 | 38.0 |
| Diversified farming operations | 27 | 29.3 |
| Sold livestock | 22 | 23.9 |
| Joined the Conservation Reserve Program | 20 | 21.7 |
| Replanted fields to forage | 13 | 14.1 |
| Cropped hay on Conservation Reserve Program lands | 13 | 14.1 |
| Refinanced or acquired new loans | 13 | 14.1 |
| Replanted fields to grain | 11 | 12.0 |
| Got out of farming | 2 | 2.2 |
| Other | 13 | 14.1 |

climate was changing, farmers most frequently cited the Greenhouse Effect or made reference to climate cycles or to a natural drying trend (Table 5.3). With regard to how climate change might affect the agricultural community, respondents cited the need for new drought-resistent crops and the importance of conserving moisture, and tended to expect decreased yields and the likelihood that many would be forced out of farming.

### Drought and Agricultural Well-Being

Perhaps one of the more curious aspects of the drought in North Dakota is how little measurable effect it had on farm financial well-being (at least in terms of gross and net income and farm debt). While North Dakota farmers did not increase their earnings significantly in 1988 or 1989, most suffered no serious financial set-backs and were able to reduce debt and increase net worth due to a variety of factors (e.g., increased land values and expectations of better prices as grain stocks were reduced). The message from 1988 and 1989 was that federal aid would continue to be available in future droughts, thereby providing an additional sense of security in farm investment.

Table 5.3
In Your Opinion, Why is the Climate Changing?
(33 respondents, 38 responses)

| Response | Freq. Cited | % Giving Answer |
|---|---|---|
| Greenhouse Effect | 12 | 27.3 |
| Part of a cycle | 7 | 15.1 |
| Naturally dryer/warmer conditions | 7 | 15.1 |
| Air pollution | 5 | 12.1 |
| Changing weather patterns | 4 | 9.1 |
| Human activities like space exploration | 3 | 6.1 |

It was probably inevitable that some critics would suggest farmers had actually benefited, perhaps unfairly, from the drought and the public assistance it evoked. A *New York Times* headline on October 16, 1988, declared: "On the Farm, a Disaster that Wasn't," and the Associated Press released an investigative series in December 1988, entitled "Cashing in on the Drought" (McCartney and Bayles 1988) based on Freedom of Information Act requests covering U.S. Department of Agriculture (USDA) entitlement reports on 12,000 farms. The study was not designed or reported in a way that allows reliable conclusions about the proper or improper distribution of federal drought aid, but it nevertheless reflects a negative public perception of farm disaster aid.

The drought also had another effect that seems counter-intuitive to the outside observer: it occurred during the most ambitious government effort to reduce the use of marginal land since the 1930s. The Conservation Reserve Program (CRP), enacted as part of the 1985 Farm Bill, was making progress in land retirement, especially by reducing dryland wheat cropping in the Great Plains. But USDA concerns over grain stocks and farmer pressure caused the agency to reduce the required set-aside by half (from 20% to 10%) for 1989, resulting in an expansion of dryland cropping in areas specifically targeted for long-term retirement for soil, water, and habitat conservation. In North Dakota, the effect of the CRP is visible as a slight decrease in wheat acreage from 1986 to 1988 (Figure 5.5) as the program began to attract farmers (the marked drop in 1983 was due

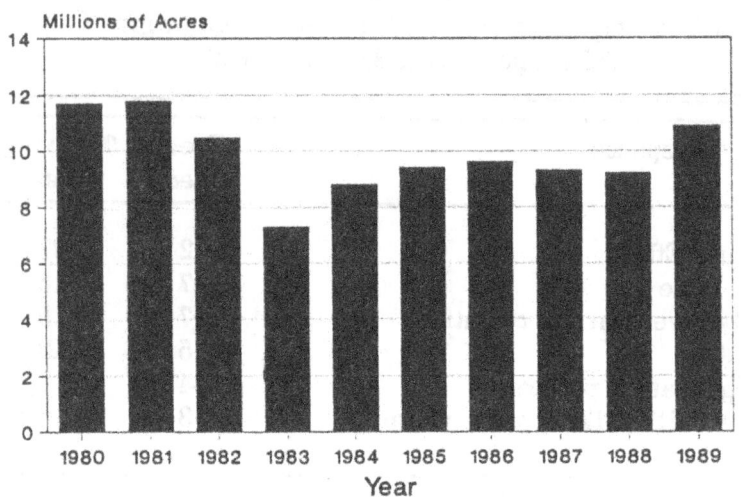

Millions of Acres

**Figure 5.5**
North Dakota wheat acreage, 1980-89 (Data from the North Dakota
Agricultural Statistics Service)

to a similar program called Payment in Kind, which was cancelled
after one year). The combined effects of reduced set-aside and higher
prices are visible in the marked increase in land cropped in 1989
(Figure 5.5). Unlike the droughts of the 1930s and 1950s, which were
followed by enhanced efforts to retire marginal lands, the 1988-89
drought brought on a reversal of conservation efforts.

## Implications for the Future

Despite its severity and negative effects during two consecutive crop
years, the 1987-89 drought in North Dakota had little deleterious effect
on overall farm financial health. It did, however, raise farmers'
concerns about the future: two-thirds of the farmers responding to a
mail survey in October 1988 reported that the drought will affect their
mid- to long-term management strategies, chiefly by making them
more risk-averse.

Many farmers were concerned that droughts are becoming more
frequent and that the hot, dry summer of 1988 might be a precursor
of changes in the region's climate associated with the much-publicized
Greenhouse Effect.

Farmers employed a wide array of methods to see them through the
drought, including both financial and natural resources adjustments.
Clearly, government aid, insurance, and drought-increased prices
reduced the drought's social impacts. Unlike past events, a drought on

today's Great Plains not only causes minimal financial loss to farmers, but can lead to expanded and more intense farming, rather than the agricultural contraction observed after past droughts. Whether such mechanisms as carry-over stocks sold at drought-inflated prices, disaster aid, and crop insurance will reduce hardships in future droughts depends on public policy and, of course, the global grain market. In 1987-89, however, these mechanisms appear to have ameliorated the effects of one of the worst northern Plains droughts of this century.

### References

Aakre, D., A. Leholm, and F.L. Leistritz. "Assessing the Severity of the 1988 Drought on North Dakota Farms and the Impact on the State's Economy." Paper presented at the Drought Water Management Workshop, November 1-2, Washington, DC. 1988.

*Bismarck Tribune.* June 18, 1988. "North Dakota has Lost 450,000 Acres of Topsoil."

Leistritz, F.L., B.L. Ekstrom, J. Wanzek, and T.L. Mortensen. *Economic Effects of the 1988 Drought in North Dakota: A 1989 Update of the Financial Conditions of Farm and Ranch Operators.* Report No. 248. Fargo, North Dakota: Department of Agricultural Economics—Agricultural Experiment Station, North Dakota State University. 1989.

McCartney, S. and F. Bayles. "Cashing in on the Drought: Some Farmers Made Profit with Drought Relief Money." Associated Press and *Boulder Daily Camera*, December 11, p. 4A. 1988.

North Dakota Agricultural Statistics Service. *Farm Reporter*, August 17. Fargo, North Dakota. 1988.

———. *Crop-Weather*, August 22. Fargo, North Dakota. 1988.

North Dakota State Water Commission. *North Dakota Water: A Reference Guide.* Bismarck, North Dakota: North Dakota State Water Commission. 1987.

# 6

# Drought-Induced Water Supply Problems at Atlanta

## Introduction

Most major metropolitan areas of the United States depend upon surface water sources (lakes, reservoirs, and rivers) for their water supply, and persistent droughts can seriously threaten the adequacy and quality of these supplies. How these threats develop and the managerial responses to drought-induced water shortages are issues of great concern in assessing the overall drought hazard in the United States.

A case of evolving water shortages in a major metropolitan area was selected for examination as part of this broader study of the 1987-89 drought. However, since droughts severe enough to affect water supplies of large cities usually last longer than one or two years, the urban water supply problems in Atlanta, Georgia, were examined because the area had been experiencing intermittent drought for several years in the 1980s.

## Background

Atlanta and its suburbs are largely supplied by water from Lake Lanier, a large, multipurpose reservoir managed by the U.S. Army Corps of Engineers (USACE). The reservoir is located approximately 35 miles northeast of Atlanta and is impounded by Buford Dam on the Chattahoochee River. Constructed in the 1950s, Lake Lanier is a source of hydroelectric power (along with other downstream reservoirs) and is also used for recreation, water supply, and navigation purposes. The Chattahoochee River flows southward from the reservoir through the Atlanta area to the Gulf of Mexico after joining the Flint and

Apalachicola rivers. The Chattahoochee-Apalachicola river system transports mainly agricultural products and supplies.

The management of water from Lake Lanier for public and industrial uses is accomplished in various ways. One northern Georgia county pumps water directly from the lake, but the Atlanta metropolitan district takes its water from the river in coordination with water releases scheduled for this purpose at the upstream dam.

## Drought Problems and Responses

Persistent drought in the southeastern United States began in 1980 and affected much of the northern Georgia area and the basin of the Chattahoochee River (Golden and Lins 1988). The period of deficient precipitation continued intermittently for several years, with a very severe dry period starting in 1985 and ending temporarily in 1986. A third period of severe drought began in the fall of 1987 and continued through 1988. Thus, the region experienced three episodes of severe drought over an eight-year period of generally below-normal precipitation, with each episode lasting one to two years. These three periods of drought are used to organize this case study of impacts and responses.

### The 1980-81 Period

In the 1980-81 drought period, state and federal officials in the Georgia area were without any drought plans, particularly with respect to the management of multipurpose reservoirs like Lake Lanier. The problems in supplying water to Atlanta were not serious, but this dry episode resulted in actions that were to be valuable in future droughts.

Primary among these was the development, at USACE District Headquarters in Mobile, Alabama, of an Interim Drought Management Plan for the Apalachicola-Chattahoochee-Flint Basin that was completed in April 1985 (U.S. Army Corps of Engineers 1985a). This plan evolved as part of a Memorandum of Agreement between the states of Alabama, Florida, Georgia, and the Mobile Corps District, and was designed to provide a long-term solution to the basin's complex water problems. The plan addressed a variety of issues, such as basin characteristics, water use and availability problems, existing drought management efforts, and institutional constraints. It also identified the need for a Drought Management Committee (DMC), consisting of USACE officials and representatives from each of the states involved, to coordinate and develop management responses from each participating agency.

The 1985 plan was quite timely, even though it only addressed drought planning on a regional level. In a broader context, it came shortly after the federal responsibility for emergency water planning was transferred from the Department of the Interior to the Army Corps of Engineers in 1983. Executive Order 11490, issued in 1969, had established an emergency water program, but droughts were overlooked in the 1985 Corps of Engineers emergency water planning program for the U.S. (U.S. Army Corps of Engineers 1985b).

## The 1985-86 Period

The second drought episode in the southeastern United States began in 1985 and continued through 1986. Precipitation and streamflow averages were below normal for much of 1985 in Georgia; precipitation was 65% of normal and streamflow was 50% of normal in the Chattahoochee Basin. These dry conditions were followed by very low rainfall in the winter and spring of 1986. Streamflow in the Chattahoochee was well below average for the first eight months of 1986, and ground water levels, which are usually highest in April and May, had fallen far below normal.

The 1986 drought set several records in the Southeast. Flows during the first nine months of 1986 established new record-low seasonal values for the Chattahoochee. The lowest flows ever observed occurred in late July. Lake Lanier set a record-low level, 16 feet below normal pool elevation, during the summer of 1986—the level was so low that boaters were cautioned to watch for submerged objects.

The Interim Drought Management Plan, available at the outset of this event, was invaluable. It outlined the overall initial strategy for coping with the 1986 drought, including the structure that gave the Corps of Engineers its general direction, but also avoided the fixed priorities and rigid commitments common to previous operational plans. Other Corps districts, either those without a plan or in the process of developing one, used it as a guide.

Water supply shortages first occurred in June 1986 at a few of the Atlanta metropolitan systems, primarily because of high demands and little backup storage. As the drought continued, several systems in the southern part of the Atlanta metropolitan area also experienced water supply problems, and several water authorities either restricted or banned outdoor water use.

In response to the worsening water shortage within the basin, the Drought Management Committee was convened and developed a plan, the Drought Water Management Strategy for the Apalachicola-Chattahoochee-Flint Basin, to manage the remaining available water stored in Lake Lanier and the other four reservoirs until the drought ended.

In August, worsening conditions at Lake Lanier led the DMC to reduce releases from Buford Dam and to serve only the water supply needs of Atlanta and its suburbs. Releases to generate hydroelectric power and to serve downstream navigational needs were curtailed or stopped. These restrictions continued until the spring of 1987, when conditions improved.

The DMC used several indicators to assess the status of the drought. These included (1) standard climatic data, (2) the Lake Lanier Water Availability Index (WAI), (3) an extended streamflow prediction (ESP) model, (4) probability of Lanier's pool elevation returning to normal, (5) total available water in storage, and (6) ground water levels.

After recovery from the severe conditions in 1987, the DMC called for a careful evaluation of how correct and useful each indicator had been during the 1985-86 period. Climatic indicators were largely based on precipitation and the Palmer Hydrological Drought Index (which was found to be an especially useful indicator). The Lake Lanier WAI, created specifically for the basin above Buford Dam, was found to fluctuate too widely to be of much use. However, calculations of the probability that Lake Lanier would return to normal pool elevation were found to be one of the more valuable drought monitoring tools. Thus, the 1985-86 drought offered an empirical test of these various indicators.

Adjustments to this drought period illustrate effective responses tied to contingency planning. The Corps of Engineers, through its water control management role, was involved from the beginning. As the drought continued through 1986 and became more severe, the Corps was increasingly called upon to provide leadership and coordination for federal and state agencies because of its reservoir operations, leadership on the Drought Management Committee, authorship of the interim drought management plan, and technical assistance program. In addition, because of its multipurpose role in managing Atlanta's water supply, the Corps worked with other levels of government and the private sector.

In assessing this experience, the Corps identified several valuable lessons for drought management (U.S. Army Corps of Engineers 1988):

1. the need for a drought contingency plan;
2. the importance of having a regional Drought Management Committee;
3. the value of timely water supply and water use data;
4. the significance of having up-to-date water control manuals and reservoir rule curves for low-flow operations;
5. the usefulness of simulation models for assessing impacts;
6. the importance of open communication and public information;

7. the need to develop Memoranda of Agreement between the Army Corps of Engineers and other institutions;
8. the relevance of having a drought monitoring and response plan; and
9. the value of coordination between Corps divisions and districts.

One interesting lesson was learned regarding citizen participation in drought planning and management. Citizen concerns over water supplies in Lake Lanier led to the formation of a local Save the Lake Committee in 1986. This group was active in requesting information from and providing information to the Drought Management Committee, thereby encouraging wider distribution of drought information than in the past. Recognizing the growing public concern, the USACE Mobile District Office held regular press briefings and sought greater dissemination of its regular lake and stream condition forecasts. The district office also began issuing statements on planned changes in lake operations and their potential impacts.

The 1986 drought also illustrated the need for reliable measures to determine the beginning, severity, and end of a drought and the need to link responses to different severities of drought. It reaffirmed the importance of up-to-date water management manuals and the use of reservoir rule curves and water control plans specifically designed for drought conditions. The regulation manual and operating rules for water control in Lake Lanier had been developed in the late 1950s and approved in April 1960 and had not been revised since, even though they were technically in force throughout the 1980-88 drought period. However, more flexible operations, under guidance from the DMC, avoided several problems.

*The 1987-88 Period*

The third severe drought period occurred in 1987-88. Extremely dry conditions returned to the area during the summer and fall of 1987. In November 1987, the Corps of Engineers, in concert with the Drought Management Committee, restricted water releases from Lake Lanier for both hydroelectric generation and navigation. Thus, only water for urban and industrial uses was being released to serve the Atlanta area. The drought indices tested and revised in the 1985-86 drought period were used to support this decision, though no formal changes in the rule curves had yet been made by the Corps of Engineers.

The impacts in the Atlanta area during 1988 were actually quite mild. All outdoor use was disallowed from noon until 9 p.m., and water use by industry was curtailed, yet no marked economic losses occurred.

However, curtailed water releases from Lake Lanier had deleterious effects on hydroelectric power generation and on downstream transportation, just as they had in 1986.

## Implications for the Future

The drought in the southeastern United States that began in 1980 and persisted through 1988 was marked by three periods of extremely dry conditions, 1980-81, 1985-86, and 1987-88. The initial severe drought period found water managers of Lake Lanier, the reservoir containing Atlanta's water supply, ill-prepared to meet the impending water shortages for various users (urban, navigation, hydropower, and recreation).

These problems led to the development by the Army Corps of Engineers and concerned states of an interim plan to better address future droughts. This plan was very useful during the severe drought episode of 1985-86. Various indices of drought severity were tested during this drought—some were improved and some were discarded. This interim plan and the establishment of a Drought Management Committee in July 1986 led to a strategic management plan during the 1988 drought that incorporated greater flexibility and integration of drought responses. In both the 1985-86 and 1987-88 droughts, a hard choice was made: water use from Lake Lanier was restricted solely to urban and industrial water needs, and water for hydropower generation and navigation was discontinued. As a result, the urban impact was limited to the bothersome, but relatively inconsequential, restriction on outdoor water use in peak hours.

The effects of the southeastern drought on the water supply of metropolitan Atlanta were not great. In fact, no great shortage of water occurred. The important lessons from the Atlanta drought can be learned from the institutional adjustments and activities taken over the eight-year period to detect, monitor, and respond to drought more effectively. Outdated regulation manuals were revised, drought contingency plans were generated, and public-sensitive strategies for water allotments were developed on a regional basis. Since such a severe drought is pervasive, everyone involved in water use needed to be involved in the process of allocating water. Analysis of these drought episodes illustrates the value of a drought plan and keeping such a plan up-to-date. In a more general sense, the Corps' reactions to the droughts in the Southeast, and primarily in the Chattahoochee basin, have provided a model for developing comparable plans elsewhere (U.S. Army Corps of Engineers 1988).

The Corps of Engineers made some very difficult decisions that involved curtailing water to many users. The lesson here is that a strategic plan, coupled with an open public process, can help water managers make such difficult choices. Through such processes, priorities can be set, hard choices can be made, and public support can be created for future drought management actions.

Another important aspect of the drought relates to long-term future concerns. Atlanta is expected to double in population by the year 2020 (Chatelain 1988). Thus, future droughts are expected to produce more severe problems than those of the 1980s, indicating a need to enhance local water supplies. The *Southeast Drought Action Report* (Chatelain 1988) identified seven federal activities needed to improve response to growing climate sensitivity, including technical and financial assistance for developing urban water supplies, development of drought management plans for all river basins, and adoption of a continuing federal program for drought.

### References

Chatelain, D.J. "Southeast Drought Action Report." In *Drought Water Management*, Proceedings of a National Workshop held in Washington, D.C. November 1-2, 1988. Edited by Neil S. Grigg and Evan C. Vlachos. Fort Collins, Colorado: Colorado State University.

Golden, H.G., and H.F. Lins. "Drought in the Southeastern United States, 1985-86." In *Drought Water Management*, Proceedings of a National Workshop held in Washington, D.C. November 1-2, 1988. Edited by Neil S. Grigg and Evan C. Vlachos. Fort Collins, Colorado: Colorado State University.

U.S. Army Corps of Engineers. *Interim Drought Management Plan for the Apalachicola-Chattahoochee-Flint River Basin*. Mobile District. 1985a

———. *Emergency Water Planning, An Evolving Program*. Washington, DC. 1985b

———. *Drought Water Management Strategy for Corps of Engineers Reservoirs in the Apalachicola-Chattahoochee-Flint Basin*. Mobile District. 1986.

———. *Lessons Learned from the 1986 Drought*. IWR Policy Study 88-PS-1. Institute for Water Resources, Water Resources Support Center, Fort Belvoir, VA. 1988.   .

# 7

---

# Drought and
# Ecosystem Management:
# The Yellowstone Fires

## Introduction

Drought during the spring and summer of 1988 induced the worst western wildfire season of recent times, continuing a trend toward larger fire areas (Figure 7.1). The fires stressed the nation's fire fighting resources and elicited a searching scientific, political, and public examination of fire policy on public lands, especially national parks and wilderness areas. The 1988 fire season earned a place in the annals of American wildfire as a time when decision makers were surprised at the unusual behavior demonstrated by ecosystems under climatic stress.

Cumulative dryness over several years, intense spring and summer drought, unusual windiness, and a century of forest evolution yielding uniform age stands converged in 1988 to produce the most severe fire conditions in the history of western wildfire management. The fires burned hotter, faster, and larger than expected, resulting in an event that took on the characteristics of a major natural disaster—a crisis often interpreted by the news media and politicians as a management mistake rather than the extreme natural event it really was.[1]

It is obvious in retrospect that "once in a century" or rarer conditions cannot be easily predicted nor controlled, and that managers cannot continually prepare for the worst case. Many of the criticisms of the so-called "let it burn" policy were misinterpretations by news reporters and political representatives (Schullery 1989). The goal of this case study is not to assess fire policy nor to analyze the capabilities of the federal fire suppression system. Rather, the focus is on the problems

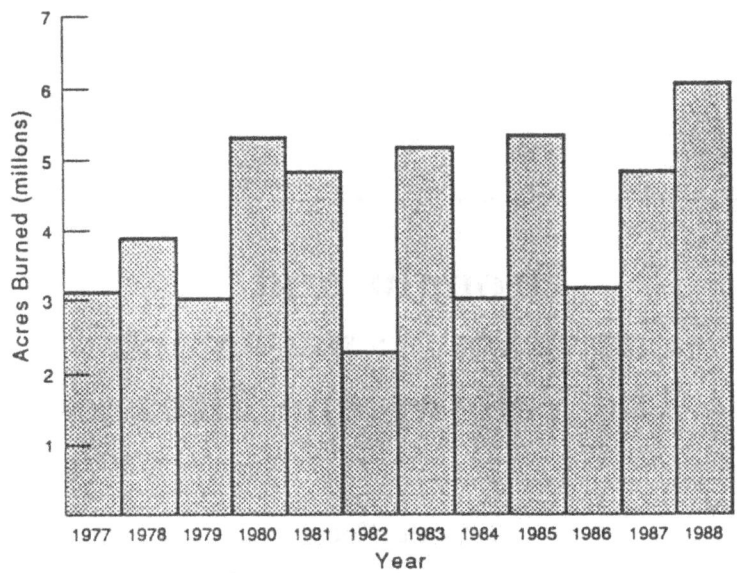

**Figure 7.1**
**Total acreage burned in forest fires, 1977-88** (*Source*: U.S. Department of
**Agriculture)**

of evaluating climate and fire conditions in cumulative drought
situations, placing the Yellowstone fires within the context of a broader
problem—the difficulties encountered when natural resource managers
must try to anticipate and respond to the behavior of complex natural
systems under unusual climate stress.

## The 1988 Fires

Wildfires occurred throughout the West in 1988, but those of the
northern Rocky Mountains in and around Yellowstone National Park
attracted the greatest public attention and will have the greatest bearing
on future fire policy. The spring and summer of 1988 were the driest
in Yellowstone National Park's 112-year history, and despite above-
normal spring rains (Figure 7.2), cumulative dryness, extreme high
temperatures, and unusual windiness combined during the summer to
produce severe burning conditions and fire behavior that surprised even
the most experienced fire analysts and fighters.

By any standard, the 1988 fire season in the Yellowstone area was
spectacular, producing the most individual fires and burning the greatest
area in the park's history. Over 50 individual fires started within the

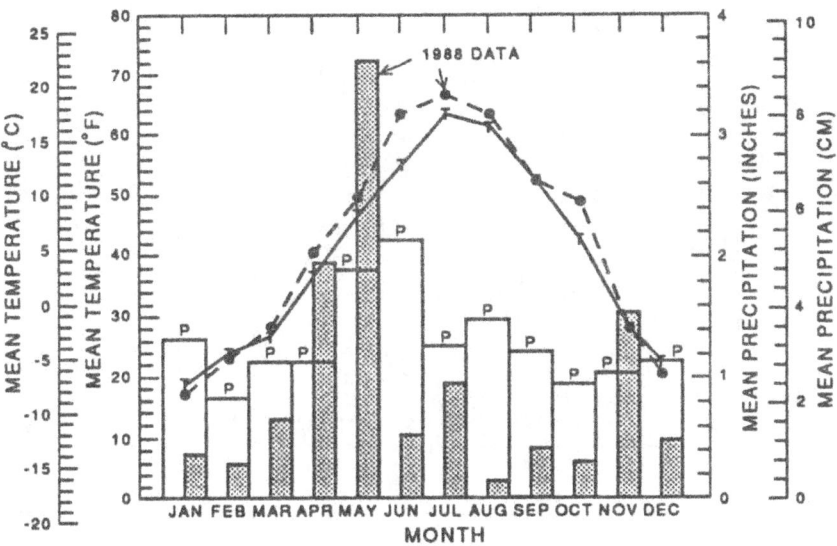

**Figure 7.2**
Temperature and precipitation for Mammoth, Wyoming, 1951-80 averages
and 1988 measurements (*Source*: Department of Atmospheric Sciences,
University of Wyoming)

park itself, while several of the larger fires, including those that
attracted the greatest attention by threatening border towns and Old
Faithful Village, moved in from adjacent Forest Service land.

Eventually, some 1.1 million acres were incorporated within fire
perimeters inside the park (Figures 7.3a and 7.3b), although, as park
officials tried (often in vain) to communicate to a public dismayed by
scenes of crowning flames carried by the nightly news, areas within fire
perimeters were not totally destroyed. Some fires burned mostly in the
substrate vegetation, leaving the tree canopy relatively unaffected, while
others jumped and "spotted" within a larger perimeter. Overall, probably
only 50% to 70% of the land within fire perimeters actually burned (Interagency Task Force 1988). High intensity burns—which produce the
conventional forest fire image of a blackened, decimated landscape—probably only occurred on 22,000 acres, or roughly 1% of the total park area.

Four major fire complexes affected the Yellowstone area (Figure 7.3b).
The North Fork fire was the single largest, covering 500,000 acres after

134

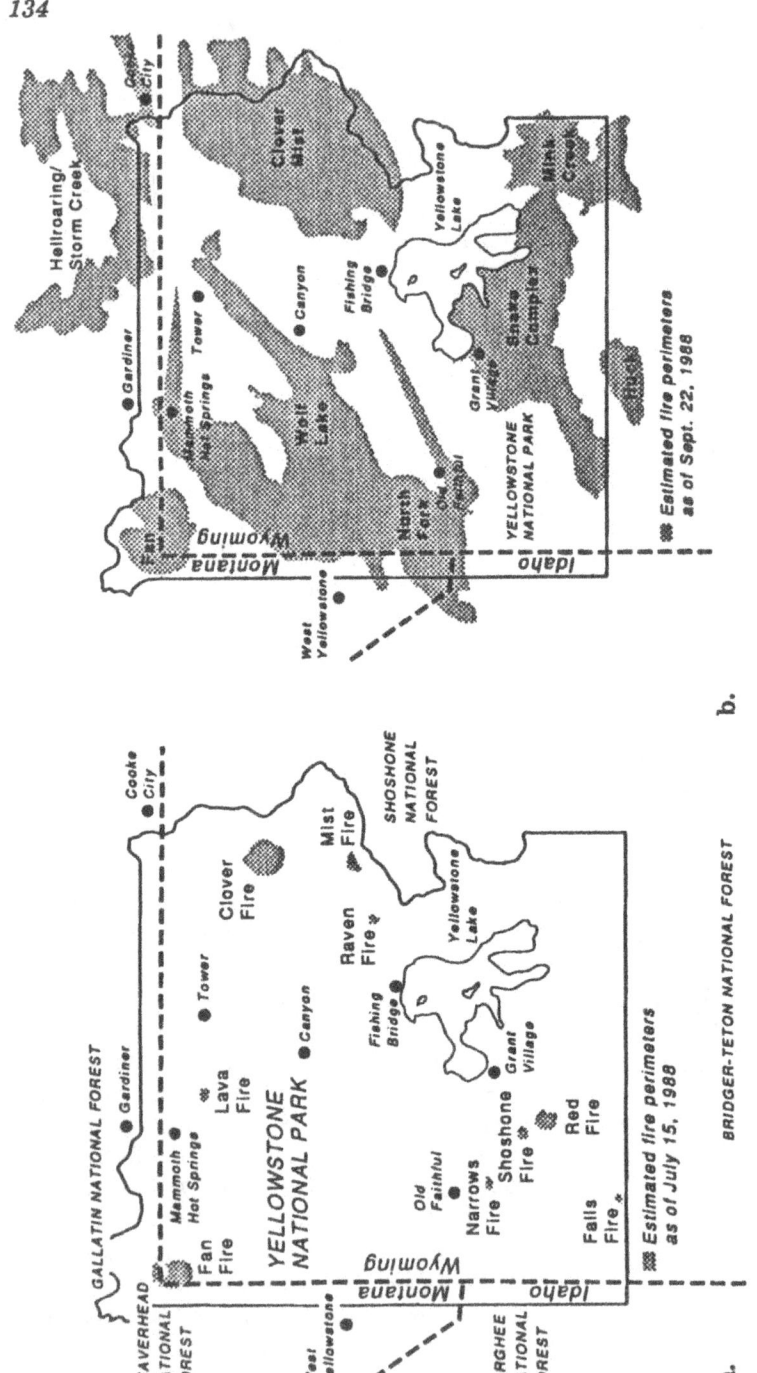

**Figure 7.3**

Estimated fire perimeters in Yellowstone park on (a) July 15, 1988, and (b) September 22, 1988 (*Source:* U.S. Department of Agriculture and National Park Service)

starting on July 22 due to accidental human ignition in the Targhee National Forest just beyond the park's western boundary. Contrary to media reports, the fire was fought from the outset both on Forest Service and park property (North Fork/Wolf Lake Fire Review Team 1988), an important point because it illustrates how 1988 fire behavior was so severe that fire fighting and control were largely ineffectual.

The North Fork fire epitomized 1988's unusual fire behavior. It spread to 460 acres in the first 24 hours and sent out spot fires as much as two miles from its front. By July 25, the fire had expanded to 2,500 acres, accelerated by dry fuel conditions and winds associated with one of several unusual dry weather fronts that swept across Yellowstone that summer. The North Fork fire later became one of the most controversial because of assertions that it could have been suppressed early, if not for the Park Service's "let it burn" policy.[2] In this setting, small decisions took on momentous proportions. For example, the fire incident commander's request for bulldozers to build a fire break within the park was initially denied in accordance with park policy and "light on the land" tactics appropriate to a natural preserve (North Fork/Wolf Lake Fire Review Team 1988). Such decisions received voluminous media attention, and fed the notion that park personnel were not actively fighting the fires. Yet, reviews show that the North Fork fire received full suppression from first attack, with little effect—nor was it ever declared a *prescribed fire*, the designation given to fires that are allowed to burn in order to maintain long-term forest health.

### Fire Behavior in 1988: How Unusual Was It?

Post-audits of the 1988 fire season produced by the federal land management agencies are unanimous in their assessment that the year was "extraordinary, unusual, unprecedented, surprising, extreme, unpredictable, unexpected," etc. All these adjectives are undeniably true—more faster-spreading fires occurred than ever before. The summer was the driest on record in the Yellowstone ecosystem, with the most severe string of negative drought indices ever recorded in Wyoming's upper Yellowstone drainage climatic division, and unusually high fire danger indicated by various measures employed by land management agencies.

The largest previous fire in park history burned only 25,000 acres in 1886—without fire suppression. Before the prescribed fire policy implemented in 1972, most fires were quickly and easily suppressed. Of the 235 prescribed fires that occurred between 1972 and 1987, 201 suppressed naturally before they grew beyond one acre (Interagency Task Force 1988). Thus, the fires of 1988 were truly unusual: fires spread at speeds of up to two miles per hour, advancing five to 10 miles daily—twice or three times the typical rate. The fires did not "lie down" at night as traditionally expected, and traveled through areas of light fuel that would normally not burn. Fires jumped landscape features traditionally thought of as solid barriers (e.g., rivers, roads, and alpine ridges) and spotted up to two miles ahead of the advancing fire front.

Under these unusual conditions, fire behavior predictions, based largely on empirical data, tended to underestimate burn rates as weather and climate factors, especially dryness and wind, exceeded the domains of experience and statistical models. According to the North Fork review team:

> Planning and execution of control actions almost became a daily exercise in futility due to the extreme, unpredictable fire behavior. Predictions were made using the best available methods of calculation and up-to-the-minute data. Rates of spread were calculated daily, only to have the fire frequently substantially exceed the predictions. The history of the fire is filled with prediction, such as "the fire will not be at point 'A' for 6 days," only to have the fire 2 miles beyond point "A" in a day and a half.
>
> A team of nationally recognized fire behavior specialists was brought in and began to make their predictions on August 1. The fire also substantially exceeded their predictions. (1988: 18)

The obvious conclusion: "Conditions and factors influencing fire behavior were in a critical state that far exceeded the capabilities of the existing predictive models" (p. 18). As a result, decision makers were repeatedly surprised at fire behavior. Even when they consciously tried to create worst-case scenarios of fire spread, and then doubled those scenarios, the fires outstripped analysts' predictions (Interagency Task Force 1988; Clover/Mist Fire Review Team 1988).

Expressions of surprise from fire commanders, longtime park managers, and even scientists who have reconstructed fire behavior over centuries in the western U.S. are too numerous to document here. The interagency review team referred simply to the "frustration and wonder" felt by fire commanders (Interagency Task Force 1988).

## Convergent Forces and the Yellowstone Fires

Multiple factors converged in 1988 to produce the most severe fire season in Yellowstone's history. First, the forest ecosystem had evolved, under natural and social forces that operate at the decadal time scale, into a state ready for extensive, hot fires. Second, cumulative drought created a climatic state in which the full fire potential of the evolving forests could be manifest. Finally, the extraordinarily hot, dry, and windy summer of 1988 provided the triggering conditions.

## Ecosystem Factors

Spring and summer dryness in 1988 pushed the Yellowstone forest ecosystem past a critical threshold where extensive, hot, and often crowning fires became the rule, rather than the exception. But the stage for severe fires was set by a combination of cumulative environmental processes, both natural and human-controlled, operating at longer time scales.

During this century the Yellowstone forests evolved into a state ripe for burning. With few large fires in historical times, most of the region's forests have matured, and the lodgepole pine stands have reached late stages of succession (Interagency Task Force 1988). Reconstructions throughout the northern Rocky Mountains show a reduction of fire frequencies during the last century (Arno 1989). One forester commented that the northern Rocky Mountain forests now comprise the oldest vegetation landscapes ever in their natural history (Williams 1989), what some have called a "climax climax" vegetation (Houston 1973). Human development inside and outside the park, especially since the 1960s, placed more property at risk than ever before, a pattern common on the fringe of most of the country's wildland areas (Gardner, et al. 1985; Gardner and Cortner 1988).

Finally, despite a prescribed burn program, little of the park had burned either naturally or intentionally for two chief reasons: (1) conditions in which the Yellowstone forests will burn are actually quite rare in this relatively moist and cool climate at high altitude (Sellers and Despain 1976; Brown 1989), and (2) prescribed burning, whether the ignition is natural or intentional, is controversial, especially in national parks, and despite stated policy, managers face public reproach when they allow fires to burn popular areas such as Yellowstone. Land managers are leery of designating prescribed natural fires that might leave their jurisdiction (though interagency agreements are in place for this) or that might be viewed negatively by the public (Lewis 1989; van Wagtendonk 1989).

This set of forces, some under natural and others under human control, set the stage for significant fires when an exceptionally dry and windy summer came along.

## Climate Factors

The Yellowstone ecosystem experienced a rapid desiccation in 1988 that was aided by a multiyear cumulative dry trend. The PDSI record shows that region experienced more dry than wet months in the 1980s,[3] but that summers were near normal through 1986 (Table 7.1). The severe dry conditions leading to the 1988 fires emerged in early 1987 and intensified in a stair-step fashion into late 1988 (Figure 7.4).

The descent into severe drought started in the winter of 1987-88, with the PDSI reaching –6.37 in March, a value unprecedented so early in the season. The lowest value, –7.11 in October 1988, was exceeded only in the summer and fall of 1977, also a large fire year in the West. The late-1970s dry spell, however, lacked the hot temperatures and strong winds of 1988.

### Anticipating the Unusual in Natural Systems Management

In these unusual conditions, fire danger assessment became quite difficult. There were some early climatic indicators of the potential severity of the 1988 fire season, and the federal agencies were gearing up for a bad season before the large fires started. While hindsight analysis cannot appreciate the difficult, often ambiguous, decision-making environment that land managers faced in the spring and summer of 1988, some aspects of western wildfire management, and the post-fire analyses by teams of land managers and researchers, suggest the need for greater attention to the use of multiple climatic data in decision-making.

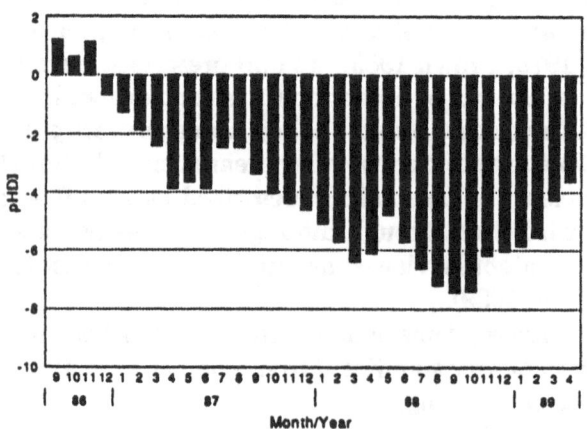

**Figure 7.4**
**Palmer Hydrological Drought Index (PHDI) for northwest Wyoming climatic division** (*Source*: National Climatic Data Center)

## Table 7.1
## Palmer Drought Severity Indices (PDSI)
## for Wyoming's Yellowstone Drainage climatic division, 1980-89

| Year | Jan. | Feb. | Mar. | Apr. | May | June | July | Aug. | Sept. | Oct. | Nov. | Dec. |
|------|------|------|------|------|------|------|------|------|------|------|------|------|
| 1980 | -4.66 | -4.74 | -4.70 | -5.22 | -4.34 | -4.46 | -3.94 | -3.27 | -2.58 | -2.95 | -2.61 | -2.39 |
| 1981 | -3.01 | -3.01 | -3.44 | -3.81 | 1.30 | 1.61 | -.33 | -1.11 | -1.74 | .10 | .02 | .19 |
| 1982 | .06 | .19 | .48 | 2.26 | 1.70 | 1.79 | 1.94 | 2.31 | 2.16 | 2.41 | 2.32 | 2.25 |
| 1983 | -.35 | -.97 | -1.28 | -1.67 | -2.15 | -2.54 | .44 | .27 | .52 | .78 | .89 | .86 |
| 1984 | -.72 | -1.57 | -2.42 | -2.62 | -2.65 | -2.83 | -2.12 | -1.82 | -1.32 | -.94 | -.85 | -.99 |
| 1985 | -1.59 | -1.87 | -2.00 | -2.72 | -3.05 | -3.49 | .16 | -.16 | .60 | .24 | .80 | .15 |
| 1986 | .26 | 1.41 | -1.14 | -1.17 | -1.88 | -2.40 | .44 | .60 | 1.24 | .64 | 1.16 | -.77 |
| 1987 | -1.41 | -1.99 | -2.55 | -3.96 | -3.77 | -3.97 | -2.39 | -2.38 | -3.25 | -3.97 | -4.29 | -4.55 |
| 1988 | -5.09 | -5.69 | -6.37 | -6.04 | -4.79 | -5.61 | -6.33 | -6.98 | -7.10 | -7.11 | -5.89 | -5.81 |
| 1989 | -5.57 | -5.24 | -4.40 | -3.82 | -2.95 | -2.96 | -2.85 | -2.43 | -2.96 | -2.41 | -2.51 | -2.46 |

## Early Indications

The 1988 fire conditions were truly unusual. Some factors, like the frequent dry frontal passages that brought strong winds and little rain to Yellowstone, could not be anticipated more than a few hours or days in advance. Moreover, monthly and seasonal outlooks from the National Weather Service did not indicate unusually dry conditions. But the overall severity of the fire season was presaged in some climate and fire-danger indices before the season got underway.

Mild drought prevailed by December 1987 and worsened quickly through the winter. Unusually severe drought was evident by March (Figure 7.4). Similarly, some early-season fire indicators were out of the ordinary and beyond the experience of fire managers involved in planning for the season. For example, the Energy Release Component (ERC) of the National Fire Danger Rating System (NFDRS) (see Deeming, et al. 1977), was higher at Mammoth, Wyoming, than ever before for early April and continued to exhibit unprecedented values through the summer—well ahead of the severe northern Rocky Mountain fire season of 1981 (Figure 7.5a). The moisture content of *thousand-hour fuels* (large tree branches and trunks that take roughly 1,000 hours or 42 days to adjust to ambient changes in humidity) was also below previously observed values in the first half of April, though it recovered somewhat during the brief wet spell in early May (Figure 7.5b).

However, other indications were ambiguous. Higher elevation stations—where the forests are—lagged behind Mammoth's valley-bottom conditions (Figures 7.6a and 7.6b), not exhibiting especially unusual indices until June. Additionally, the National Weather Service's seasonal forecasts for the summer, issued in late spring, called for near-normal temperatures and precipitation. Many of the six-to-14-day forecasts in late June and July were for hot and dry conditions, but these could not anticipate the strong winds that propelled the fires. The 30- and 90-day forecasts that might have presaged the season never suggested especially dry conditions.

## Problems in Interpreting Drought and Fire Indicators

In this difficult-decision setting, fire analysts were called on to make critical projections and choices involving the allocation of short resources and the management of park and other public lands. Fire narratives (e.g., Carrier 1989) recounted and examined these decisions, as did official review teams. The examination was sometimes painful, often politically sensitive, and trying for the individuals with decision-making responsibility during the fires. Yellowstone's superintendent protested especially the manner in which fire policy was simplified, sensationalized, and

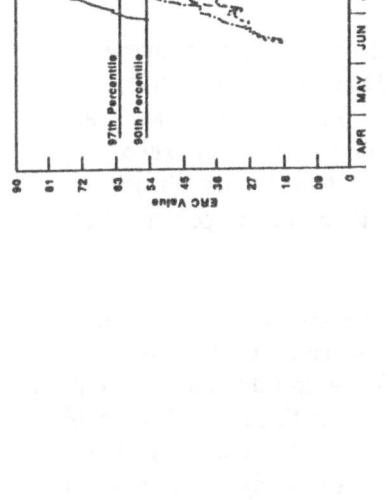

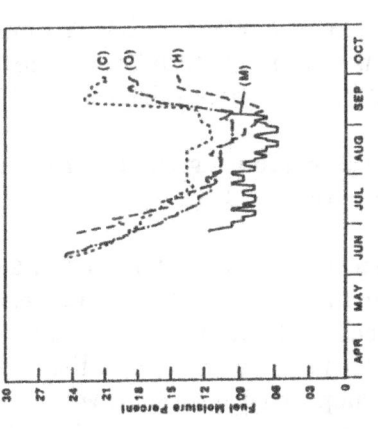

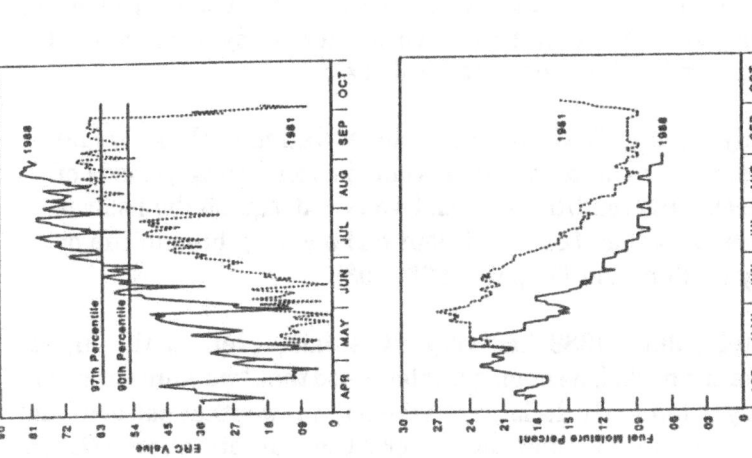

**Figure 7.5**

(a) Energy Release Component (ERC) and (b) 1,000-hour fuel moisture for Mammoth, Wyoming, 1988 (*Source:* U.S. Department of Agriculture)

**Figure 7.6**

(a) Energy Release Component (ERC) and (b) 1,000-hour fuel moisture for selected Yellowstone sites, 1988. M=Mount Sheridan, H=Mount Holmes, O=Old Faithful, and C=Canyon.

misinterpreted by the news media (Barbee 1989), but the federal review panels were also critical. The Clover/Mist Fire Review Team summed up the complex decision-making situation faced by land managers early in 1988 well:

> Managers recognized individual pre-season indicators, but information was not cumulatively interpreted to recognize the full potential for a severe fire season. (1988: 6)

Difficulty in evaluating the severity of fire conditions continued into the second week of July:

> Fire specialists did not feel there was a problem of an abnormally severe fire season. The Clover and Mist fires were declared "Prescribed Natural Fires" July 9 and 11 after analysis was conducted by a team of specialists. At this time, [managers] could have suppressed both fires with initial action forces. Projected spread and worst case analysis underestimated the potential of these fires (1988: 5-6).

The review team indicated that fire managers relied too much on history as their guide in 1988, despite indications that an historically unique fire season was underway. For example, based on past experience, managers expected the majority of fires burning in May to suppress naturally in June, which is the wettest month at many Yellowstone sites (Martner 1986). An earlier study that supported increased prescribed burning concluded that:

> Naturally caused fires [in Yellowstone National Park] would have plenty of room to run their course even during periods of extreme burning conditions ... and would extinguish themselves as naturally as they began, without endangering human life or property (Sellers and Despain 1976: 103).

Unfortunately, June 1988 had very little rain, even in the higher elevations of the park. One manager later noted that "we were definitely basing many of our early decisions on the normal weather pattern" and that the unusual weather had not yet "grabbed [our] attention" (*Denver Post* October 16, 1988).

Problems arose in integrating and interpreting multiple climate and fire indicators. Yellowstone analysts, quoted in the October 16, 1988, *Denver Post*, said they tended to focus their attention on the 1,000-hour fuel moisture index rather than the Palmer Drought Index or

Energy Release Component. Given their "mind-set that, based on the last seven years, we'd have a mild winter followed by a cool, moist summer," the 1,000-hour fuel index was not sufficiently alarming to evoke expectations of truly unusual conditions.

According to the federal review team, managers were, early on, overconfident about their ability to control fires and tended to underestimate fire spread rates because they extrapolated past behavior to the 1988 season. Thus, several northern Rocky Mountain fires were initially declared prescribed burns. Some fires burned under these conditions into July, when all fires were classified as *wildfires* requiring suppression. Postseason evaluations, however, indicated that essentially all fires in the region were "out of prescription" by either intuitive or quantitative measures much earlier (Fire Management Policy Review Team 1988).

A more subtle problem arose from managers' perceptions of how flammable the forests were. Previous studies had shown that large sections of the Yellowstone ecosystem were naturally ready to burn (Sellers and Despain 1976; Despain 1977). Furthermore, many of the Northern Rocky Mountain forests exhibited a fire behavior profile marked by a climatic threshold beyond which fires burn hotter and spread faster than is the typical case (Figure 7.7). Past some climatic threshold not easily distinguished by the fire danger rating system, and given sufficient wind, fires would readily spread by crowning, even in relatively young forest stands (Rothermel 1989). This nonlinear behavior makes modeling and intuitive or experienced-based fire assessment especially difficult.

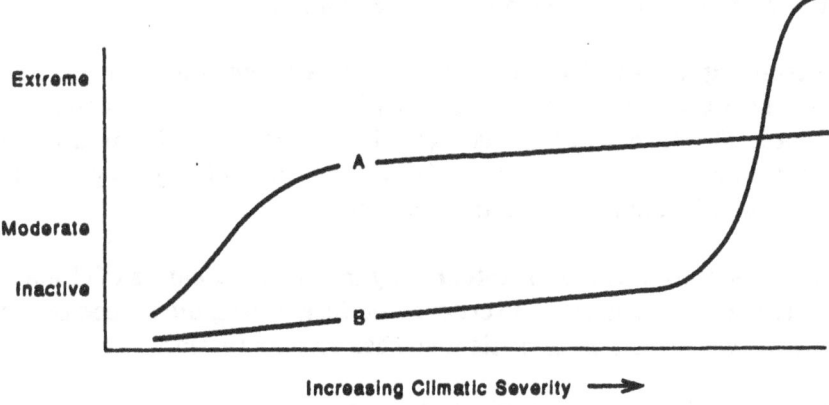

**Figure 7.7**
"Threshold" behavior of fires in (a) middle-successional and (b) late-successional forests (*Source:* Williams 1989)

## Several Factors Combined to Yield Surprise

In summary, several factors contributed to the difficulty of assessing fire behavior during 1988.

1. Little in the historical record or the two decades of detailed data in the NFDRS foreshadowed the potential for extremely dry and windy conditions in the region, especially in the higher elevations, though prehistoric fire reconstructions showed that large, region-wide fires are not unprecedented. Historical factors, like past fire suppression and natural forest evolution, combined to create a landscape that was ready to burn. These processes operate on long time scales ill-suited to short-term analysis and the policy-making process.

2. Fire prediction models are largely empirical (i.e., based on past statistical relationships between climate, weather, fuel, and fire behavior) rather than "process" (e.g., the physics of fuels combustion). As in the case of a regression model forced to predict conditions well outside the domain of empirical data from which it was derived, the predictions were typically underestimates.

3. According to interagency review teams, fire behavior exceeded the experience and the imagination of most frontline decision makers. Even the prudent safeguard of calculating quantitative or intuitive worst-case scenarios was inadequate: the 1988 fires, driven by unusually strong winds and spreading through the crowns of drought-desiccated trees, exceeded these extrapolations.

4. According to some fire behavior analysts, and outside observers and reviewers, fire managers tended early in the season to count on "normal" weather and climate conditions even though cumulative dryness had set the stage for intense fires, given triggering events such as midsummer hot and windy spells.

5. Methods were lacking for integrating multiple indicators of climate stress and relating these to criteria for designating prescribed burns versus wildfires and to critical choices on suppression.

With rapidly deepening drought and unusual windiness, even the most pessimistic fire analyst was bound to underestimate the potential for fires to burn large areas quickly during 1988. Surprise was inevitable under such conditions, but some thought on how to interpret and integrate climate data, especially early in unusually dry seasons, might reduce some future surprises.

## Reducing Future Surprises in Managing
## Climate-Sensitive Natural Resource Systems

The drought and fire season of 1988 can be viewed as a case of a complex environmental system behaving under stress in a manner that frustrates intuition, experience, and technical analysis. Holling (1986) introduced the idea of surprise in several studies, arguing that resource planners tend to expect uniformity and normality of natural processes, often neglecting the potential for unusual conditions to produce rare or counterintuitive interactions between policies and ecosystem behavior (e.g., pest control and forest health).

The few studies of resource manager response to conditions as unusual as those in Yellowstone during 1988 tend to support this notion. In his analysis of the rapid rise of the Great Salt Lake in 1983, Morrisette (1988) found that lake managers acted on limited notions of lake level behavior because they expected conditions to tend toward "normality" such that unusual conditions would either be short-lived or compensated by future conditions.

Another case that closely parallels the Yellowstone situation is the unprecedented runoff in the upper Colorado Basin during 1983 (see Rhodes, et al. 1984). Discharge of the Colorado River exceeded all previous records that spring, and water managers were not prepared for the rapid runoff and rise of reservoir levels. Unusually deep snowpack was apparent all winter and well into the spring, and managers were aware that an intense runoff was possible. An unusual May heat wave, however, acted as a triggering event that, with abnormally deep snowpack and saturated conditions from several seasons of above-normal precipitation, caused the catastrophic melt and runoff. Unfortunately, the Colorado River reservoirs were managed as in a normal year, in which the goal is to store as much water as possible, and emergency spilling, causing flooding and damage to control works, was necessary.

## Are Surprises Inevitable?

Recognizing the problems inherent in anticipating truly unusual conditions, what can be learned from the 1988 fire season and other cases to improve future management of climate-sensitive natural resources? The many fire management reviews have produced long rosters of specific lessons, and this case study offers a few more general examples.

First, fire prediction models based on historical data and normal climatic conditions are bound to underestimate truly unusual conditions. Thus, models based on the physics of combustion and fire processes should receive more attention. Statistical models will be calibrated with the 1988 conditions to extend their range, and work is underway on models for persistently crowning fires. Effort is also needed to explore methods for evaluating early-season indicators and integrating multiple indices. Although many indicators (e.g., 100-hour fuel moisture, Palmer Hydrological Drought Index, and the Energy Release Component) indicated severe conditions early in the 1988 season, no widely accepted formulas beyond rules of thumb were available to assess how these indices reflected longer-term conditions that would not be ameliorated by a switch to normal, or even above-normal, rainfall.

These and other lessons and policy proposals appear in detail in the many reviews that emerged after the fires were out (e.g., Fire Management Policy Review Team 1988), but the more significant, and general, lesson to be learned from 1988 is how managers of complex natural systems deal with unusual conditions that result from cumulative processes operating at different time scales.

The 1988 experience shows that even worst-case analysis may not help managers anticipate truly rare events and that the potential for catastrophic resource system behavior under climate stress may be underestimated, even with evidence that unusual or unprecedented conditions are developing. Thus, managers should pay more attention to cumulative trends that set up potential catastrophes (long-term system evolution and persistent climate anomalies) and link this with analysis of the potential "wild cards" or triggering events that could flip the system into a different behavior pattern. This may be especially difficult if short-term swings, such as the wet conditions in late-May and early-June, distract attention from underlying forces that are pushing ecosystems toward thresholds of new behaviors.

Drought in particular may be difficult to integrate with natural resources planning and management. The progression of a major drought may be asymmetric due to complex interactions of precipitation, runoff, infiltration, and evapotranspiration in natural and man-made systems. That is, a return to "normal" conditions will not necessarily end the drought, and the precipitation needed to reduce established drought conditions may be a greater proportion of normal than was the deficit that initiated the drought.

Perhaps the most important lesson from 1988 is the need to assess the potential for triggering events on top of severe cumulative conditions. The risk of unusual windiness or the absence of crucial rains during typical wet periods must be analyzed in addition to traditional factors that worsen the fire hazard.

## Implications for the Future

The fire season of 1988 was a success for many land and resource managers. A well-planned and efficient fire-fighting system was tested and shown to be in good shape. While budget cuts have reduced Forest Service and Park Service fire-suppression capacity, the agencies were still able, through mutual aid agreements, local/state cooperation, and military assistance, to mount the largest fire management response in the country's history, one costing roughly $500 million. Few injuries or deaths occurred, and very few homes or other developments were affected. Park rangers and foresters argue that the ecological benefits of the fires outweigh their costs, despite media and public perceptions that the fires "destroyed" large areas.

It is unlikely that earlier suppression would have made a large difference in the eventual extent of the 1988 fires. Rather, the case tells us something about anticipating and responding to convergent forces that can make ecosystems behave in an unusual manner and points out the need to incorporate the full range of available climate information in decision-making and to have *a priori* decision-frame-works in place—and agreed upon by managers—for dealing with indications of especially unusual and severe conditions. This is particu-larly important in terms of western wildfires, because many of the West's forests remain ripe for high-intensity burning. But, it also applies to other natural resources management settings where long- and short-term climatic trends, interacting with processes inherent to natural and social systems, can yield conditions that cannot be predicted using traditional tools.

### Notes

1.  An informal survey of news media coverage of the 1988 fires shows that public lands managers were regularly portrayed has having let the fires get "out of hand" (*New York Times* September 22, 1988; *Chicago Tribune* November 20, 1988), having failed to anticipate and respond to extreme fire danger despite ample evidence that the 1988 season would be severe (*Denver Post* October 16, 1988), and even covering up their bad decisions and adherence to misguided principles of natural forest regulation (*Wall Street Journal* August 26, 1988). While many of these criticisms can be seen in retrospect to be myths and over-blown sensationalism, it is clear that the agencies did not deal well with the media's quite predictable tendencies toward controversy and their focus on issues, such as restrictions on the use of fire-fighting equipment in protected areas,

which were trivial in terms of overall fire strategy, but evocative of public concern. It is worth noting that the agencies received some positive and sympathetic coverage of their fire preparations in early July before the truly spectacular fires were burning (*New York Times* July 10, 1988).

2. "Let it burn" or "let burn" is the popular and somewhat misleading term for prescribed fire policies now in place on most federal lands. Prescribed fires are those allowed to burn under certain criteria based on weather, the feasibility of control, and the threat posed to natural or human values. "Prescribed natural fires" are those started by natural forces (e.g., lightning) while planned human ignitions are called simply "prescribed fires." Other fires are termed "wildfires" and subjected to various levels of suppression, and a prescribed fire may be reclassified as a wildfire if conditions change.

3. The Palmer indices reported here are calculated for Wyoming's Yellowstone drainage climatic division, which includes most of the park. A combination of low and high elevation stations is included in the calculations.

## References

Arno, S.F. "Fires of 1988 Viewed in a Long-Term Perspective." Paper presented at the Wilderness/Parks Fire Conference, May 23-26. Montana State University, Bozeman. 1989.

Barbee, R. Foreward. In J. Carrier, *Summer of Fire*. Salt Lake City: Gibbs-Smith. 1989

Brown, J.K. "Could the 1988 Fires Have Been Avoided with a Program of Prescribed Burning?" Paper presented at the Wilderness/Parks Fire Conference, May 23-26. Montana State University, Bozeman. 1989.

Carrier, J. *Summer of Fire*. Salt Lake City: Gibbs-Smith. 1989

*Chicago Tribune*, November 20, 1988. "Let-It-Burn Policy Put Heat on Officials."

Clover/Mist Fire Review Team. *Clover/Mist Fire Review*. Washington, D.C.: U.S. Department of Agriculture. 1988.

Deeming, J.E., R.E. Burgan, and J.D. Cohen. *The National Fire-Danger Rating System-1978*. General Technical Report INT-39. Ogden, Utah: U.S. Forest Service, Intermountain Forest and Range Experiment Station. 1977.

*Denver Post*, October 16, 1988. "Yellowstone Ignored Two Key Fire Indicators."

Despain, D.G. *Forest Succession Stages in Yellowstone National Park*. Information Paper No. 19. Wyoming: Yellowstone National Park. 1977.

Fire Management Policy Review Team. *Report on Fire Management Policy*. Washington, D.C.: U.S. Departments of Agriculture and Interior. 1988.

Gardner, P.D., and H. Cortner. "When the Government Steps In: Making Public Policies to Regulate the Wildland/Urban Interface." *Fire Journal* (May/June): 32-37. 1988.

Gardner, P.D., H. Cortner, and J.A. Bridges. "Wildfire: Managing the Hazard in Urbanizing Areas." *Journal of Soil and Water Conservation* 40: 319-321. 1985.

Holling, C.S. "The Resilience of Terrestrial Ecosystems: Local Surprise and Global Change." In *Sustainable Development of the Biosphere*, edited by W.C.

Clark and T.E. Munn. Cambridge, United Kingdom: Cambridge University Press. 1986.

Houston, D.B. "Wildfires in Northern Yellowstone National Park." *Ecology* 54: 1112-1117. 1973.

Interagency Task Force. *Greater Yellowstone Area Fires of 1988: Phase II Report.* Washington, D.C.: U.S. Departments of Agriculture and Interior. 1988.

Lewis, E. "Conservationist Perspective." Paper presented at the Wilderness/Parks Fire Conference, May 23-26. Montana State University, Bozeman. 1989.

Martner, B.E. *Wyoming Climate Atlas.* Lincoln: University of Nebraska Press. 1986.

Morrisette, P.M. "The Stability Bias and Adjustment to Climatic Variability: The Case of the Rising Level of the Great Salt Lake." *Applied Geography* 8: 171-89. 1988.

*New York Times*, July 10, 1988. "Amid Drought and the Heat, U.S. Braces for Rash of Fires.

——, September 22, 1988. "Ethic of Protecting Land Fueled Yellowstone Fires."

North Fork/Wolf Lake Fires Review Team. *Review of the North Fork/Wolf Lake Fires.* Washington, D.C.: U.S. Department of Agriculture. 1988.

Rhodes, S.L., D. Ely, and J.A. Dracup. "Climate and the Colorado River: The Limits of Management." *Bulletin of the American Meteorological Society* 65: 682-91. 1984.

Rothermel, R.C. "Yellowstone Park Long-Range Fire Predictions in 1988." Paper presented at the Wilderness/Parks Fire Conference, May 23-26. Montana State University, Bozeman. 1989.

Schullery, P. "The Story Itself: Lessons and Hopes from the Yellowstone Fire Media Event." *The George Wright Forum* 6 (3): 17-25. 1989.

Sellers, R.E. and D.G. Despain. "Fire Management in Yellowstone National Park." In *Proceedings of Tall Timbers Fire Ecology Conference.* Washington, D.C.: U.S. Department of Agriculture. 1976.

van Wagtendonk, J.W. "Variations in Policy Implementation." Paper presented at the Wilderness/Parks Fire Conference, May 23-26. Montana State University, Bozeman. 1989.

*Wall Street Journal*, August 26, 1988. "Yellowstone Burns as Park Managers Play Politics."

Williams, J.T. "The Case For/Against Planned Ignitions in Wilderness." Paper presented at the Wilderness/Parks Fire Conference, May 23-26. Montana State University, Bozeman. 1989.

# 8

## Managing Drought
## in the U.S.:
## Problems and Solutions

Overall, how well did the nation manage the widespread drought
that began in 1987, worsened in 1988, and continued into 1989? In
this chapter, we offer our assessment of how government agencies and
other key groups responded. We then offer a set of planning recommendations that will better prepare the country for future droughts.

### Major Governmental Responses

Government decision makers treated the drought as an emergency
requiring extraordinary, short-term measures to support farmers, protect
resources, fight fires, assist communities with water problems, and
provide relief in the manner of response to other types of natural
disasters. Although routine mechanisms came into play, emergency
responses received the greatest attention and, indeed, were generally
effective at reducing the drought's most drastic social impacts through
loss compensation. The weaknesses in governmental response appeared
in terms of prehazard planning and mitigation, long-term drought
vulnerability reduction, and monitoring and anticipation, as discussed
in the next section.

The principal agency affected by the drought was the U.S. Department
of Agriculture (USDA). Its disaster assistance programs were changed
and elaborated under emergency conditions, including special assistance
to livestock producers because of poor hay crops and grazing conditions,

temporary permission for haying and grazing on land withdrawn from production under mandated set-aside and conservation programs, and, of course, a package of financial relief measures. These extraordinary measures were added to standing hazard management programs like all-weather peril insurance, administered by the semiprivate Federal Crop Insurance Corporation, which insured 45,000,000 acres (21% of all farm acreage) and paid out roughly $3 million in 1988.

The Forest Service and National Park Service mounted the largest fire-fighting effort in history, including assistance from the military. The National Park Service in particular found itself operating in a crisis mode, facing more fires than ever before and attracting severe public criticism for its handling of the Yellowstone fires (see Chapter 7). In retrospect, it is clear that 1988 fire conditions were truly extraordinary, and, despite late recognition of the full severity of fire potential, it is doubtful that fire suppression efforts could have been any more effective than they were. In sum, surprisingly few properties were damaged by the fires, and ecologists count the burning of Yellowstone as an environmental gain, rather than loss, from the drought.

Other federal agencies involved in water resources and water management undertook emergency drought activities during 1988. The U.S. Army Corps of Engineers (USACE) operated 12 dredges in the lower Mississippi to keep that river open, and the Coast Guard restricted barge and tow sizes. The Fish and Wildlife Service took on several special projects to protect wildlife refuges and other habitats and to maintain minimum stream flows, especially in the West. The Tennessee Valley Authority (TVA) adjusted reservoir system management to meet various multipurpose demands for scarce water. Due to reduced river flows, the TVA spent $105 million to purchase power in 1988, and the Southeastern Power Administration purchased $30 million in supplemental electricity. The Corps of Engineers constructed an underwater sill in the Mississippi River at New Orleans to halt saltwater intrusion up the river (due to low flows) and to protect the water supply of New Orleans.

The Corps of Engineers also took a variety of temporary steps to regulate their multipurpose reservoirs in the Southeast to assure minimum water supply and quality under the cumulative stress of runoff deficits. This entailed, however, reduced water service for hydropower and recreation (General Accounting Office 1989). The agency funded the costs of transporting water and drilling new wells in certain drought areas. The Bureau of Reclamation established drought coordination teams in each of its regions, rescheduled water deliveries, provided technical assistance for weather modification programs in the

West, and delivered some water beyond its normal service areas under emergency authorization.

There were notable instances of effective interagency drought response in 1988 involving multiple federal, state, and local agencies. The drought often created regional problems that required responses from all levels of government (e.g., around the Great Lakes). Successful efforts to protect Atlanta's water supply through effective management of Lake Lanier and the Chattahoochee River system occurred because of regional and multilevel government planning and coordination (Chapter 6). However, these efforts succeeded primarily because two drought episodes during the 1980-86 period of persistent southeastern dryness provided the experience that evoked cooperative management in time for the 1988 drought. Another example of "getting the players together" to improve drought response, and one that involved the private sector, is the interaction of the Corps of Engineers, the Coast Guard, and barge owners in response to low Mississippi River levels. This was an ad hoc reaction, but serves as an exemplar of planned interaction. Further examples are described in Grigg and Vlachos (1990).

The National Oceanic and Atmospheric Administration (NOAA) experienced a surge in demand for its data and analyses. NOAA personnel filled increasing numbers of information requests during the spring (e.g., to farmers and fire-danger analysts) through routine channels (e.g., existing links between Forest Service and NOAA meteorologists). By mid-June, however, the agency's Climate Analysis Center (CAC) began issuing special drought reports and providing situation briefings by drawing on its own data sources and ancillary information. The CAC's 30- and 90-day forecasts were in high demand, despite the low accuracy with which they are able to foretell precipitation trends. NOAA's National Climatic Data Center (NCDC) in Asheville, North Carolina, provided much of the climatological analysis for the special reports. NOAA also coordinated with the U.S. Department of Agriculture (USDA) through the Joint Agricultural Weather Facility. Once these systems were operating under "drought alert" conditions, they utilized effective interagency cooperation to provide invaluable information on a time frame useful to resource managers. Finally, additional research was supported by the National Science Foundation (NSF) to study drought impacts and management, for example, through the Natural and Man-Made Hazards Program in engineering (e.g., Grigg and Vlachos 1990).

Effects on the federal government can also be measured in relation to the federal budget. The agricultural relief alone ultimately amounted to over $7 billion paid to farmers through subsidized insurance and direct payments, though this was partly compensated by reduced expenditures on income supports and price subsidies. However, this may be a one-time option, and cannot be counted on to reduce the economic effects of future droughts as crop subsidy policy changes.

About a third of the states affected by the drought had previously undertaken some form of drought contingency planning that was helpful in establishing state responses in terms of monitoring, assistance, and coordinating with federal programs. A few states provided outright financial relief to drought victims via guaranteed loans and reduced interest rates. Most states aided water-short communities with technical assistance, especially for conservation programs, while some provided direct aid in the form of state-built pipelines from more reliable water sources and the drilling of new wells. State National Guard units were employed to haul water, enlarge rural ponds, and make other minor physical improvements in water supply systems. Several states altered regulations for water use and water release from municipal- and state-controlled reservoirs in cooperation with federal agencies.

The drought also evoked extensive legislative activities in Congress and statehouses around the country. At least a dozen different bills relating to drought relief were introduced in Congress. By mid-July, a bipartisan agricultural relief bill had been prepared, and the Disaster Assistance Act of 1988 was signed into law on August 11. This act chiefly provided assistance to farmers and led to the expenditure of $4 billion in relief by May 1989. Continuing drought in 1989 evoked further action, including the 1989 Drought Relief Act that provided $900 million to ameliorate further drought impacts.

## Problems of Adjustment and Response

While short-term reaction to the 1987-89 drought appears to have been successful in averting disaster in the traditional sense, many of the government agencies responsible for managing climate-sensitive resources were ill-prepared and rather slowly realized the drought's full potential to disrupt resource systems and critical flows of goods and services. Hence, the agencies reacted late and at times incorrectly. The drought's severity and apparent sudden onset caught many resource managers off-guard. Plans for assessing and responding to drought conditions were lacking or were poorly implemented early in the drought. Where contingency plans did exist, they made a difference (as in the case of Atlanta's water supply), yet many were out of date and hence inappropriate to the situation (as in the Mississippi River system). Drought responses in 1988 frequently took on the characteristics of crisis management, such as the call for emergency water diversion from the Great Lakes or calls for immediate changes in fire-fighting policy.

*Problems in Drought Anticipation and Monitoring*

The crisis nature of government response to the drought was at least partly due to tardy attention to the emerging hazard. Some agencies began to recognize the drought in the spring and began at least to consider what to do if it worsened, but most agencies did not begin to respond in substantive ways until late June or early July, despite the record dry spring weather. The drought was already severe in the spring, and was truly intense by mid-June. It was not until then that a coordinated federal response began in earnest. In the last weeks of June, President Reagan established the Interagency Drought Policy Committee and Congress established the Drought Relief Task Force. The Department of Agriculture established a national hotline to disseminate drought information. NOAA also instituted a drought hot line, began special drought briefings in mid-June, and began to issue special drought advisories on June 22. Yet, earlier indicators of emerging severe drought were not used to issue early warnings, and interests that might have benefited from early warnings in the spring received little information or advice.

The April-June 1988 period was one of record drought intensification in the U.S. The climatological analyses in Chapter 2 show that drought was well underway in the spring and evident in some places during the winter. Drought indices disclosed the emergence of serious late-winter and early-spring dryness in essentially all areas destined to be affected by severe summer drought in 1988. A quarter of the country was experiencing severe to extreme drought by May. Several areas, such as the northern Great Plains and Rocky Mountains, exhibited measurable drought conditions in late 1987, while much of the Southeast had exhibited drought conditions since 1984.

Yet, analysis and distribution of climatological and water resource data were not placed on a drought-alert basis until mid- to late-June, well after the severity of the drought was apparent to most observers. Valuable information outlets, like drought situation reports from NOAA, were not available until even casual observers, and of course the news media, could see for themselves the physical effects of drought reflected in croplands, wetlands, and forests.

Some early indicators of serious drought problems, such as rapidly falling river levels on the Mississippi in April and May and unusually low fuel moisture values during April in the western forests, were not given the credence they deserved. Perhaps resource managers hoped that the dryness would not continue into the summer, or perhaps they were waiting for more concrete impacts to avoid "crying wolf." In the case of a cumulative hazard like drought, however, once the impacts are evident—when physical and ecological systems have "dried out"—*above-normal* precipitation is required for timely recovery even to *normal* moisture levels. Moreover, by the time impacts are obvious, drought is well entrenched and it is too late for many resource managers to take mitigating steps.

Resource managers noticing early indications of drought may also have been waiting for some form of *institutional declaration* that a drought existed, an official and public recognition of the hazard that would validate their concern over drought indicators and emerging impacts. This validation problem is less severe in response to obvious, rapid-onset hazards like hurricanes, earthquakes, and floods, suggesting the need for a different approach to drought monitoring and impacts assessment—one that makes the drought threat more tangible before obvious impacts occur.

The case of spring fire danger assessment and, indeed, of overall drought impact assessment, also suggests a bias toward optimism, even in the face of complex environmental processes with potentially large negative consequences. Managers often expect systems to respond to environmental conditions as they have in the past, even though those systems change through time (e.g., forests growing older) and may respond in a nonlinear manner to cumulative stress. Managers of complex environmental systems tend to expect past patterns to repeat themselves, even in unusual periods—a tendency evident in the fall of 1988 when pronouncements were made that 1989 would be a better year because the country had never experienced two back-to-back severe drought years. In combination, these factors delayed hazard recognition and response.

*Some Key Impacts Neglected*

Drought is a pervasive hazard, affecting most sectors of the economy and other aspects of society. However, drought responses in 1987-89 tended to focus on some impacts while neglecting others. For example, the USDA primarily aimed its drought programs of information, advice, and financial aid toward farmers and livestock producers, leaving agribusinesses to fend for themselves. Many businesses were unprepared for drought and made dramatic changes in production quotas, marketing plans, procurement, and sales using whatever information they could garner from private and public sources. Some larger agribusiness firms established internal drought task forces and altered corporate plans in midyear to cope with the drought. But, few firms were aware of the Interagency Drought Policy Committee's economic impact assessments, which were issued monthly after midsummer, but not widely distributed.

Another neglected sector was public health, especially with regard to heat stress and the mental health effects of drought. Much was learned about the health consequences of heat and drought in 1980, yet response in 1988 was similar to that in 1980: no warning to vulnerable populations, little government information or education efforts at any level, and an uncoordinated mixture of emergency responses (extra medical staff and ambulance patrols, hastily designated "cooling centers") in urban areas where the most concentrated vulnerable populations—the poor, elderly, and ill—live. While urban areas like Kansas City and Dallas/Ft. Worth did apply lessons from 1980, there was no widespread cautionary information dissemination and little national impact assessment.

Heat-wave deaths are now crudely estimated from heat-stroke records and rough estimates of "excess deaths" occurring during heat episodes. We know only that somewhere between 5,000 and 10,000 people—mostly poor elderly in substandard housing without air conditioning—died from heat-related causes in 1988, and that at least some vulnerable populations were protected by emergency responses such as municipal cooling centers and neighborhood contact programs. More focused heat-hazard responses should be developed in the major metropoli, where assistance programs for low-income and elderly residents should be devised, and the capacity of current cooling systems, as well as the efficacy of building codes in reducing stress, should be better assessed.

## Drought Response Problems
## Illuminated by the 1987-89 Event

Drought is an ancient natural hazard, but in the context of a rapidly evolving technological society, the 1987-89 drought uncovered some disturbing problems in social response to natural hazards that may have been less obvious during past events. Three particular issues—(1) scientific uncertainty and confusion about the nature and cause of the drought, (2) the role of the media, and (3) the prospect of increased societal sensitivity to extreme climate events—deserve special attention.

### Scientific Confusion

The delay of information concerning the onset and severity of the drought in 1988 reflected atmospheric scientists' limited knowledge of droughts and their impacts, as well as weaknesses in our systems for monitoring and disseminating information about climate impacts. Scientists tended to focus on debates over the causes of the drought, rather than on reliable assessments of its impacts. These debates clouded the drought picture and projected a muddled image of climate and the potential for climate change through the Greenhouse Effect. Pronouncements that the drought was an early signal of global warming, and subsequent refutations of these claims, confused public and official views of the drought hazard.

This is illustrated in the recent media coverage of an analysis by Trenberth, et al. (1988) showing that the unusual circulation patterns that caused the drought were themselves caused by pre-existing sea-surface temperature anomalies. This analysis was widely reported as having proved that the drought was not caused by the Greenhouse Effect (see *New York Times* January 3, 1989). The banner *New York Times* statement that the "Greenhouse effect was not the culprit this time" implies a false initial premise: that the Greenhouse Effect could be unambiguously linked to a singular event like the 1987-89 drought. Trenberth, et al., state that their study is of the synoptic and air-sea interactive causes of the drought at the seasonal time scale, and that it does not address longer-term changes in climate. We fear that this fact was lost on the public and on many decision makers who may have to respond both to future droughts and to the cumulative effects of global warming in the years ahead.

Nevertheless, the link made between drought and global warming in 1988 could have a lasting influence on U.S. policy relating to climate change. As the North Dakota case study showed, at least some decision makers believe that the drought was part of long-term climate change. Are they then to expect droughts to be more frequent? Decision makers searching for advice on this issue will be disappointed, as current estimates suggest that an unambiguous signal of global warming—if it does occur—will not be evident for at least another decade.

The greenhouse connection aside, atmospheric scientists did a poor job of informing the country about the drought itself. Those who made statements about its severity were not careful to state the region to which their data referred or what measures of drought magnitude and extent they were using. The public heard conflicting statements about whether this was, in fact, the worst drought ever to affect the country. Our own analysis in Chapter 2 illustrates the difficulties and nuances of different drought measures and their interpretations; in some parts of the country it was, indeed, the worst on record—but not everywhere.

Scientific uncertainty and confusion were magnified by the manner in which information was conveyed, as when local and regional drought assessments received national, and even international, dissemination by news media. Late in the summer and into the fall of 1988, scientists were also quoted variously as declaring the drought "over," "continuing," or (un)likely to redevelop in 1989 without providing the basis for their assessments or conveying the uncertainty of their predictions—messages that confused decision makers worrying about 1989 drought prospects.

Moreover, statements that back-to-back severe droughts have never occurred in history and thus 1989 probably would not be a drought year were misleading. Actually, as shown in Chapter 2, U.S. droughts tend to occur in "clumps," or episodes of multiple dry seasons or years. While truly severe drought years may not have occurred consecutively in the 100-year record available for analysis, two of the worst drought years—1934 and 1936—were separated by only one *relatively* wet year. There was little scientific basis for official confidence, expressed in late 1988, that 1989 would not be a drought year (see Chapter 4). In sum, uncertainty and misinformation reflected poor scientific and lay understanding of the drought phenomenon.

*The Effect of the Media*

The drought of 1987-89 was both a climatological and media event. Media drought coverage during 1988 was probably the greatest of any year since the 1930s, and the media attention endured longer and focused on more fundamental issues (physical cause, social vulnerability, policy) than is common for other natural hazards. News media attention also flavored government responses and public perceptions of those responses. This interaction was most starkly manifest in the case of the Yellowstone wildfires, a staple of the evening news for several weeks. The park's superintendent later complained that the media grossly misrepresented the nature of Yellowstone wildfire management (Barbee 1989).

The media also made the drought more concrete for people, with omnipresent scenes of withered crops and farmers sifting dry soil through their hands. Reporters habitually compared the drought to the 1930s Dust Bowl, frequently claiming that 1988 was as bad or worse—which it was in some areas—though rarely offering climatological indices or impact measures to support their assertions (e.g., *New York Times* August 7, 1989).

*Increased Social Sensitivity*

At the national level, the 1987-89 drought was not the most climatologically severe or persistent in U.S. history, but it probably caused in one year the most drought-related impacts of any previous dry spell. Its monetary impacts rank with the potential worst-case hurricane or even a severe earthquake. The question implied by this, of course, is whether modern American society is becoming more sensitive to climate impacts, despite technological advances in agriculture, forestry, and other natural resources areas that should lower our vulnerability.

There is no easy answer to this question. On the one hand, the physical impacts suggest continuing, perhaps heightened, drought vulnerability—when it does not rain, crops die, water runs short, forests burn, and rivers dry up. On the other hand, the country had sufficient financial resources to ameliorate the worst drought impacts and was able to call on large food stocks to reduce the consumer impacts of this most recent dry spell. Elaborate water control systems, like those in the Missouri Basin and Southeast, were able to ameliorate impacts on water supply and navigation.

We submit, however, that the country's underlying sensitivity to drought—in terms of impacts on crop yields, water supply, environmental quality, and health—remains as large as ever and has probably increased over the past few decades.

In one sense, the impacts of droughts and other natural hazards are destined to increase as the country grows, as White and Haas (1975) pointed out following the first national assessment of natural hazards in the mid-1970s. More property and people are placed at risk as the country develops, especially with increasing use of floodplains, coastal zones, and seismic areas. However, it could be argued that unless hazard impacts begin to consume larger portions of gross national production or outweigh the benefits accrued from hazard-sensitive activities (e.g., living in the floodplain or farming on the Great Plains), increased loss is a necessary and acceptable ancillary to economic development. Unfortunately, data on hazard impacts—especially a pervasive hazard like drought—do not allow us to determine impact trends in any precise way, except to say that they are increasing.

We can deduce from broader trends, however, that drought vulnerability is increasing due to factors such as increased demand for water resources, an aging population, degrading public infrastructure, loss of wetlands (which act as natural reservoirs), and development that limits aquifer recharge. There is also evidence that agricultural technology aimed at producing larger average yields does not necessarily reduce the *variability* of yields; thus, yield declines during droughts have not lessened with added technology (see, for example, Warrick 1984).

Increased vulnerability comes not only from development, but from narrowing ranges of alternative adjustments. More water systems are near peak capacity now than in the past; infrastructural development has not kept up with use in many parts of the nation; and institutional rigidity in some cases reduces response options in the face of drought and other hazards, as in the case of entrenched and competing forces focused on the Great Lakes and western U.S. water.

Concerns over Great Lakes levels made the proposal to divert water into the Mississippi River system institutionally untenable, even though it may have been feasible in a technical sense. States in the Colorado Basin are close to appropriating their entire legal shares of Colorado River water, despite evidence that full allocation will cause shortages even in "normal" years. There is, in these two cases, evidence of institutional rigidity, which has the effect of reducing the options available in droughts.

Land use also plays a role in drought vulnerability. Urban development is consuming farmland and reducing the area available for groundwater recharge. In many parts of the country, decisions about wildfire management are based not as much on goals of long-term forest health, but on the need to protect development at the wildland/urban fringe (see Chapter 7). While this may not have been a major issue in the Yellowstone fires of 1988, it was important in several fires elsewhere on public lands. Economic development around forest and park lands lessens opportunities to manage these lands so that environmental processes such as fire, often necessary to an ecosystem's health, can operate in a natural way.

Finally, it should be noted that some of the factors that reduced impacts in 1987-89 were temporary and incidental. For example, existing food stocks and government agricultural subsidies helped reduce the eventual economic and social burden of the drought, but these safeguards were incidental and fortuitous and may not be available for the next drought.

## Recommendations for Improved Drought Management

What are the implications of our diagnosis of drought management capacity in the U.S.? Basically, public institutions reacted to the drought in a crisis management mode because of a lack of strategic planning for drought. Beyond specific operational plans (e.g., USACE plans for low-flow management of reservoirs and fire-fighting plans), federal agencies lacked guidelines for dealing with drought, relying instead on ad hoc responses. Some agencies, such as the Bureau of Reclamation, had to await emergency legislation to take some useful actions, such as delivering water outside their jurisdictions. This is in sharp contrast to the more comprehensive planning and delegation of authority undertaken for earthquakes, floods, and hurricanes. Yet, as 1988 illustrated, a severe drought can be as costly and disruptive as these other hazards. Some states, like California, had elaborate drought contingency plans and were better able to react, but in general, the drought also caught state and local governments ill-prepared.

In the absence of plans that link drought monitoring and response at multiple governmental levels, there was a lack of timely action during the spring and early summer of 1988 as the drought intensified nationwide. Slow transfer of information to, and interpretation by, resource managers was part of the problem. Drought information was often not shared between agencies. Even when definitive drought information was available, decision makers were hesitant to react in a manner commensurate with the drought's serious dimensions. Lack of experience with droughts, particularly severe droughts, apparently left many managers unable to conceive of the consequences of a continued dry spell. The drought highlighted the difficulties of managing natural resources in a fluctuating climate, but it also pointed to some potential solutions and needed improvements.

Several key issues must be addressed if we are to reduce society's vulnerability to climate fluctuations like the 1987-89 drought and to improve natural resources management in the face of climate uncertainty. In this section, we identify some of the problem areas and recommend solutions. Our solutions are not, of course, perfect nor all-inclusive and should be considered proposals to be improved upon and modified.

### Recommendation One: Assess the 1987-89 Experience

First and foremost, the nation's lack of general drought contingency planning at the federal level should be corrected. We feel that some of the 1987-89 losses could have been averted by a more organized national monitoring and information dissemination system. There was also little knowledge of a full range of possible responses and little *a priori* adjustment planning, as illustrated by the questionable proposal for Great Lakes water diversion. A first step toward filling this gap is for the relevant agencies to:

**Conduct detailed postaudits of 1987-89 drought experiences, and use these studies for strategic planning of future management options. Postaudits can point to a broader range of solutions, as well as actions to avoid in the future. Where evaluations of past droughts were available in 1988, managers did a better job of responding to severe drought conditions.**

Certainly these postaudits will find areas of both strength and weakness in response. For example, the Corps of Engineers (1989) surveyed field reports to compile a roster of 1988 drought responses. While offering little in the way of substantive evaluation, several lessons emerged. It was found that drought communication was all bottom-up within the agency and that field offices could have been better informed if headquarters had sent periodic reports back down the organizational structure. The main goal of such postaudits is not to allocate blame, but rather to act as the basis for improved future planning.

*Recommendation Two:*
*Create a Better Drought Watch System*

Information flow to government and private decision makers was slow, and the information too ill-defined or poorly analyzed, to allow anticipatory action, especially early in the drought. This is an enduring problem in bureaucracies where several units are given separate, sometimes conflicting and overlapping, responsibilities. In particular, we have partitioned institutional responsibilities for water resource management with little regard for the natural whole of river basins and the linkages necessary for ecologically sound and comprehensive planning. Arbitrary disaggregation of natural systems and agency myopia led to some curious responses in 1988. For instance, the president's drought task force was initially established *without* a NOAA representative.

Institutional weaknesses are especially evident when the nation faces a hazard like drought, because climate pervades and links most natural resources and economic sectors. Issues that cut across institutional lines, like water policy and climate impacts, tend to be neglected or poorly managed simply because they are not the sole purview of one agency. Furthermore, meteorologists, hydrologists, agriculturalists, and others, both government and private, offered differing interpretations of what was happening. This confused people, and revealed the lack of mechanisms for coordinating critical information. Individual agencies with bits of information needed to create a wholistic view of the situation (e.g., streamflow, snow pack, soil moisture, recent rainfall, and related impacts), but had poor mechanisms for sharing and integrating the data. By the same token, drought offers a potential bridge between agencies and interests, and at a minimum we see the need to:

**Create a well-coordinated "drought watch" program that links federal, state, and local resources agencies on a *continuous* basis and provides a clearinghouse function**

for data and information that must be translated and disseminated to decision makers.

Elements of a drought watch system are already in place, but need to be better coordinated around clarified agency responsibilities. The nation's climate information gathering and analysis system—including NOAA's National Weather Service and National Environmental Satellite Data and Information Service, the USDA's Joint Agricultural Weather Facility and fire weather services, and the USGS and USACE water-monitoring system—should work with other agencies to develop a coordinated national drought watch program that matches existing focuses on flood and severe storm programs. This should be linked to a strong regional program anchored by the Regional Climate Centers established under the National Climate Program. While there certainly will be other voices and sources of information, both public and private, a centralized clearinghouse system for climate monitoring and analysis that can be relied upon in all cases is needed in view of growing concern over climate change. Better links with the media—more frequent and fuller briefings—will further assure improved drought information dissemination.

*Recommendation Three: Improve Use of Drought Indices*

In 1988 there was an obvious problem with defining drought, determining how severe it was or might become, and then taking the appropriate action. This was *not* due to a lack of drought indicators. Rather, most indices did well in tracking the evolving drought. The problem emerged mostly in interpretation and use of the indices. Thus, we recommend:

An in-depth evaluation and improvement of the reliability and usefulness of drought indices and other climatological data for decision making and natural resources management.

A few drought analysts have compared and critiqued drought definitions and indices (e.g., Wilhite and Glantz 1985; Wilhite and Easterling 1987), but widely accepted standards for, and limitations on, their application have not been developed. Lacking guidelines, people use various drought indices indiscriminately, often producing misleading hazard assessments. The choice of a drought indicator is subject to the goal of the assessor, as illustrated in our own work in Chapter 2. Different measures of drought have different sensitivities to changes in climate conditions (e.g., the Palmer Drought Severity Index versus the Palmer Hydrological Drought Index), making them more or less appropriate for different uses (e.g., assessing the threat to dryland crops versus stress on groundwater supplies). Other parameters, like national rainfall anomalies, may be misleading unless one understands fully what conditions they reflect (e.g., moisture anomalies in normally dry versus normally wet regions).

Perhaps more important than fine-tuning the indicators themselves is the need to examine and improve their *use*. Because major droughts are rare, most resource managers rarely find themselves working with drought indices or basing critical decisions on them. Experience can be gained through *post hoc* studies of past drought periods and analysis of indicators and decisions as possible scenarios of future situations. Backcasting decisions to previous episodes and comparing actual impacts will demonstrate the best indices and how they can be applied to decision making.

*Recommendation Four:*
*Develop and Apply Climate Impact Models*

In concert with problems in assessing and acting upon drought severity, there existed an uneven ability among resource managers to project impacts on crops, forest fire behavior, water levels, and wildlife. We failed to have climate impact models running and accessible, despite the fact that much has been learned about the links between climate, nature, and society in the past two decades and that reliable impact assessment techniques have been developed for most resource sectors (see Kates, et al. 1985). It is time for the practical, real-time application of this knowledge through:

An integrated program of impact assessment and projection in which crop-yield, fire-behavior, water-resources, climate-health, energy, economy, and other simulation models are used early on to provide ranges of possible impacts based on past, existing, and potential climate patterns.

We also need to know more about the drought sensitivity of various natural systems. For example, a common question in 1988 was whether various grain crops could be revived by renewed rainfall or whether they were past the point of survival. This critical point in a crop's development varies from place to place and crop to crop, but it is a basic piece of information that should be available to all drought watchers and agricultural planners.

The tools to conduct real-time impact assessments that can guide responses and mitigation efforts should be evaluated, collected, improved, and applied in an programmatic manner. Climate impact assessment is, by nature, a diverse and interdisciplinary activity, and it would be difficult and expensive to create a centralized impacts assessment system. However, a network of impacts modelers could be established by building on the interest already focused on longer-term climate change impacts. The network would function through coordinated data management, assessment, and information dissemination and should include government, university, and private sector researchers who can bring their models of climate impacts to bear in a useful time frame of evolving drought conditions. This should become part of the existing climate information system and the improved drought watch system recommended above. Further development of drought impact assessment methods should be aided by NSF-sponsored research, especially in Atmospheric Sciences and the Natural and Man-Made Hazards Program in the engineering division, which has recently supported several useful drought studies.

*Recommendation Five: New Contingency Plans*

Some resource managers may argue that a summer such as 1988 cannot be planned for and can only be managed by emergency response. Orthodox contingency planning is relatively easy, but establishing plans that incorporate potential surprises and counterintuitive situations, and that transcend traditional responses, requires imagination and open-mindedness. Yet, unusual events are precisely the phenomena for which institutions must plan the most carefully. Less severe, more frequent stresses can be managed by routine responses and flexibilities, but extreme events can lead to wholesale system failures. We were lucky that the 1987-89 drought did not offer several years of extreme dryness over most of the country—a scenario within the realm of climatological potential. Managers also know that complex environmental systems can exhibit behavior that transcends intuition and the normative models used for routine management and planning.

Yet, we found that some important aspects of contingency planning, such as efforts to assess and expand the range of alternatives available to decision makers in the face of truly unusual conditions, are often neglected despite the fact that such planning is relatively easy, cheap, educational, and consumes no inordinate amount of staff time. Thus, we need to:

**Create more guidance on how resource managers can conduct effective contingency planning, especially for rare/extreme events and unusual climate conditions requiring nontraditional responses.**

The USACE Planning Division is currently analyzing national water management options in drought, and its report will be a useful component of a more integrated national plan. Some resource activities (e.g., reservoir management) better lend themselves to planning for unusual conditions than others, and it may be useful to explore the transferability of planning approaches between resource activities. Each area (e.g., water supply, crop marketing, fire management) has its own standards for contingency planning, but some issues are cross-cutting. For instance, a common problem is failure to consider less orthodox responses—to be satisfied with tradition. Another problem is determining how often contingency plans should be updated. This depends on the sensitivity of the resource activity and the potential for changes in the broader social and physical environments that alter the system's vulnerability. Changes that occur on long time frames—like land-use conversion, population growth, or changes in global food markets—will eventually make existing plans outdated. Researchers who study decision making and formal risk analysis should be able to provide improved guidelines on the creation of richer, risk-based plans and to recommend the rate and nature of plan revisions. Finally, federal agencies should examine the nation's ability to manage a severe, persistent drought, perhaps along the lines of USACE emergency water planning.

*Recommendation Six:*
*Increase Attention to Neglected Sectors*

Certain impact areas were neglected in 1987-89, two of which deserve special attention. We did a poor job of anticipating, monitoring, and mitigating the impacts of drought and heat on both physical and mental health. Despite rough estimates that up to 10,000 deaths might have been triggered by the drought and accompanying heat, there was little information or advice from agencies such as Health and Human Services or the National Centers for Disease Control. Yet, with its rapidly aging population and a sizeable homeless population, the U.S. must intelligently and aggressively address heat-stress hazards. We need to know more about the magnitude of the problem, provide educational programs to inform people about heat-related illnesses, and devise the means to assist people unable to cope with temporarily high utility costs.

The federal government should improve heat mortality and morbidity record-keeping, conduct studies of heat-related disease, and establish guidelines for addressing this worsening problem. Additionally, the effects of drought, heat, and other climatic stresses on mental health—a virtually unexplored problem—should be investigated.

*Recommendation Seven:*
*Improve Information Dissemination*

Another gap in governmental responses to the drought was the lack of information directed toward the business sector. U.S. businesses would be helped by better access to drought impact information generated by task forces and agencies. While it has been the goal of recent administrations to restrict public services, especially where they may compete with private enterprise, a large proportion of drought assessment and information dissemination activities logically reside with government, and few private sources can marshall the data and analysis capabilities necessary to inform decision makers about emerging hazards like drought. The agribusiness sector was especially poorly served by both public and private information sources.

The federal government should examine ways to better inform business and industrial sectors about drought impacts, and should especially consider assessing whether some rules and regulations of agribusiness might hinder drought responses.

Past efforts to involve industry in research and applications of long-range forecasts under the National Climate Program offer a model for such an effort.

## A Unifying Proposal:
## Full Implementation of the
## National Climate Program Act

The drought of 1987-89 revealed that the nation is quite sensitive to climatic extremes that affect water and other natural resources. This is due to narrowing margins between resources supply and demand, increasing population and development in sensitive environments, greater demand for environmental quality, narrowed response options, and more tightly coupled resources management systems. The recommendations above suggest some ways to reduce this vulnerability.

Taken together, our suggestions coalesce into a key observation and suggestion. Many of the efforts suggested here were also recommended, implicitly or explicitly, in the National Climate Program Act of 1978 (NCPA). Yet, provisions in the act calling for increased agency coordination for research and applications regarding climate impacts have not been implemented. Our final recommendation then is simply that the NCPA be fully implemented.

Indeed, the first element called for by Congress in creating the National Climate Program (NCP) was:

assessments of the effects of climate on the natural environment, agricultural production, energy supply and demand, land and water resources, transportation, human health and national security. (PL 95-367, 92 Stat. 601)

The country is more sensitive to climate impacts now than when this legislation was written, because certain elements of the NCPA never came to fruition. The act called for coordinated basic research and attempts to solve practical problems in climate and society interactions. It also mandated improved flow of climate data and information services for applications useful to society. While basic climate research continues, and while some data problems (e.g., barriers to international exchange) were solved, the fundamental problems of climate impacts on society were largely neglected. The National Climate Program was under-funded and spent its energy on issues other than climate impacts, such as the quest for better seasonal forecasts. These efforts did little to reduce the country's sensitivity to climate fluctuations. This programmatic gap will continue to plague us in the future, whether we face natural climate fluctuations or climate warming due to human activity.

A panel reviewing the National Climate Program, under the auspices of the National Academy of Sciences, in 1986 recognized that the NCP had neglected climate fluctuations and their impacts and concluded specifically that:

It is time to institute a coordinated interagency and inter-governmental effort to reduce our vulnerability to drought (includ-ing shortages in agricultural, urban, and industrial water supplies). Policy planning analyses of governmental response in the 1970s point to poorly coordinated reaction to, and inefficient monitoring and assessment of drought . . . The apparent lack of drought planning at the federal level suggests that the next major U.S. drought will again evoke an inefficient and poorly coordinated response, as did the droughts of the 1970s.

The [National Climate Program Office] is in a unique position to initiate planning efforts to improve [drought response] (pp. 14-15).

But, participants in the review also noted that "the topic of drought is a gap in the National Climate Program" (p. 27) and now, more than a decade after the National Climate Program Act was passed, the federal government has not fully met its responsibilities in research and information dissemination on the interactions of climate and society.

Despite a decade-long national climate program, ad hoc drought task forces, and burgeoning public attention to climate impact issues, governmental activities in climate analysis, information dissemination, and impact assessment are inadequate, poorly coordinated, and ill-prepared for increasing social vulnerability and the inevitable climate fluctuations to come. The goal now should be to fill these gaps, using the 1987-89 drought as a compelling argument for concerted action to improve the quality of climate services in the U.S. that incorporate improved monitoring, data and information management, interagency coordination, and efforts to measure the multiple interactions between climate, natural resources, and society. Ultimately, our goal must be to understand, anticipate, and modify the complex relationship between climate fluctuation and societal vulnerability.

### References

Barbee, R. Foreward to J. Carrier, *Summer of Fire*. Salt Lake City: Gibbs-Smith. 1989

Grigg, N.S. and E.C. Vlachos. *Drought Water Management.* Report to the Natural and Man-Made Hazards Mitigation Program of NSF. Fort Collins, Colorado: Colorado State University. 1990.

Kates, R.W., ed. *Climate Impacts Assessment: Studies of the Interaction of Climate and Society.* New York: John Wiley and Sons. 1985.

*New York Times.* July 8, 1988. "1988 Evokes Ghost of Dust Bowl."

——, January 3, 1989. "Greenhouse Effect Was Not the Culprit This Time."

Trenberth, K.E., G.W. Branstator, and P.A. Arking. "Origins of the 1988 North American Drought." *Science* 242: 1640-1645. 1988.

U.S. Army Corps of Engineers. *Surviving the 1988 Drought.* Paper 89-P-2, Ft. Belvoir: Institute for Water Resources Support Center. 1989.

Warrick, R.A. "The Possible Impacts on Wheat Production of a Recurrence of the 1930s Drought in the U.S. Great Plains." *Climatic Change* 6:5-26. 1984.

Wilhite, D.A., and M.H. Glantz. "Understanding the Drought Phenomenon: The Role of Definitions. *Water International* 10: 111-120. 1985.

Wilhite, D.A., W.E. Easterling, and D.A. Wood, eds. *Planning for Drought: Toward a Reduction of Societal Vulnerability.* Boulder, Colorado: Westview Press. 1987.

White, G.F., and J.E. Haas. *Assessment of Research on Natural Hazards.* Cambridge: MIT Press. 1975.

# Index